Essential Concepts of Land Politics

This book compiles, discusses, and explains the key concepts in land politics in an easy-to-navigate A–Z format.

This book takes a broad view on land, across the rural and urban corridor, and advocates for a holistic view of the politics of land, as an aggregation of land and global social life, that is, the politics of food, climate, labour, citizenship, and geopolitics. The authors have curated a wide-ranging list of 67 key terms most commonly used in the field, with each entry mapping out an important concept or idea and illustrating how it relates more broadly across this growing discipline. Across the entries, the book showcases that land has been and remains central to productive and social reproductive activities of humanity for the incessant renewal of life and society. A key assumption in this book is that the politics of land is made up of building blocks in the form of key concepts. These key concepts evolve, both in the sense that they are politically contested and in the changing broader context within which they are embedded. The key concepts in this book are therefore not discussed in a random way but rather framed from Critical Agrarian Studies perspectives and scholar-activist tradition, which means taking the side of the exploited and oppressed.

With further reading recommendations included alongside the entries, this innovative and accessible volume will be of great interest to students, scholars, policy practitioners, and political activists.

Saturnino M. Borras Jr. is a Professor of Agrarian Studies at the International Institute of Social Studies (ISS) in The Hague and is part of the Erasmus Chairs program of Erasmus University Rotterdam, the Netherlands. He is also an associate at the Transnational Institute (TNI) and a Distinguished Professor at the China Agricultural University in Beijing. He was the Editor-In-Chief of the *Journal of Peasant Studies* for 15 years, until 2023. He is the co-editor of *The Oxford Handbook of Land Politics* (2025).

Jennifer C. Franco is a researcher at the Transnational Institute (TNI) Myanmar-In-Focus programme, the Netherlands and an Adjunct Professor at the China Agricultural University in Beijing. She is the co-author of *Scholar-Activism and Land Struggles* (2023) and co-editor of *The Oxford Handbook of Land Politics* (2025).

A–Z Guides for Environment and Sustainability

This series provides accessible, easy-to-navigate overviews of a range of different topics related to environment and sustainability. Following an A–Z format, each book contains entries which map out an important concept or term and illustrates how it connects more broadly to other ideas and disciplines. With related terms and further reading included alongside the entries, these innovative volumes will be of great interest to students and scholars learning, teaching and researching in this field.

Essential Concepts of Sustainable Finance
An A–Z Guide
Edited by Elisa Aracil and Ibrahim Sancak

Essential Concepts of Land Politics
An A–Z Guide
By Saturnino M. Borras Jr. and Jennifer C. Franco

For more information about this series, please visit: https://www.routledge.com/A-Z-Guides-for-Environment-and-Sustainability/book-series/AZES

Essential Concepts of Land Politics

An A–Z Guide

Saturnino M. Borras Jr. and Jennifer C. Franco

Routledge
Taylor & Francis Group
LONDON AND NEW YORK

earthscan
from Routledge

Designed cover image: Cover image is from the paintings of Federico 'Boy' Dominguez of the Philippines. The graphic design of the cover is by John B. Tan

First published 2025
by Routledge
4 Park Square, Milton Park, Abingdon, Oxon OX14 4RN

and by Routledge
605 Third Avenue, New York, NY 10158

Routledge is an imprint of the Taylor & Francis Group, an informa business

British Library Cataloguing-in-Publication Data
A catalogue record for this book is available from the British Library

ISBN: 978-1-032-65845-2 (hbk)
ISBN: 978-1-032-65841-4 (pbk)
ISBN: 978-1-032-65843-8 (ebk)

DOI: 10.4324/9781032658438

The Open Access version of this book was funded by Erasmus University Rotterdam.

For Product Safety Concerns and Information please contact our EU representative GPSR@taylorandfrancis.com. Taylor & Francis Verlag GmbH, Kaufingerstraße 24, 80331 München, Germany.

Contents

Preface

This book is a collection of key concepts on the politics of land. The definitions and meanings we offer for the entries in this book are not technical nor are they politically neutral; rather, they are politically biased. Our ideas are from broadly Critical Agrarian Studies perspectives (Edelman and Wolford 2017; Akram-Lodhi et al. 2021; Borras 2023; Shattuck et al. 2023) and are framed from a scholar-activist tradition (Aguiar et al. 2023; Borras and Franco 2023), which means taking the side of the exploited and oppressed.

This book takes a broad view on land, across the rural and urban corridor, although it has a heavier emphasis on the rural. It is a book that makes a coherent and original argument as to how best to understand the politics of land. It can be seen from two perspectives. On the one hand, it is a book that advocates for a holistic view of the politics of land, as an aggregation of land and global social life, and argues that this view can be achieved by bringing selected keywords together so that the sum is greater than its parts. On the other hand, it is a book that offers selected key concepts related to land that are described and examined individually, leading in turn to a bigger and more systematic understanding of land politics. In short, this book argues that a big concept such as land politics can only be fully understood when seen from its disaggregated parts, and at the same time, while each of these parts stands on its own as a separate object of scientific inquiry, they can only be fully understood when seen from the totality that they constitute together.

The main argument of this book is that land has been and remains central to productive and social reproductive activities of humanity for the sustenance of life. Land has multiple meanings to diverse social classes and groups linked through various social relations. How land is controlled, by whom, through what institutional mechanisms, for how long and for what purposes, are some of the questions that animate fierce political contestations around the issue of land control. Such questions have increased in intensity and extent since the contemporary global land rush kicked in around two decades ago. If our starting point is social justice and ecological sustainability, the most urgent questions become how can these be achieved through what kind of land politics? Globally, land remains highly concentrated in the hands of a few dominant classes and social groups, and the state. Land politics must be democratized, but how can this be done in the current context? We address these questions by putting together a description of selected keywords, each of which is written in a way that shows how contested the meaning of that term or concept is in reality.

A key assumption in this book is that the politics of land is made up of building blocks in the form of key concepts. These key concepts evolve, both in the sense that they are politically contested and in the changing broader context within which they are embedded. For example, the meaning of 'land reform' has evolved and been contested through time from one society to another. The key concepts in this book are therefore not discussed in a random way but rather framed from a political economy perspective of contestation over land control.

We locate this book in the fields of Critical Agrarian Studies. In this context, land forms a thread in multiple closely related themes, each of which has its own field of studies. More specifically, a broad guiding intellectual and political question behind the framing of this book is how land politics have shaped and been shaped by different spheres of social life, that is, the politics of food, climate, labour and citizenship, as well as geopolitics. We aim to address this question by using conceptual building blocks in the form of keywords in the context of recent political developments, especially the contemporary global land rush. By looking at the parts, we hope to see the whole; by looking at the whole, we hope to reach a deeper understanding of the parts.

In order to deepen and broaden our understanding of the politics of land, this book takes as its immediate context the recent global land rush. The land rush has put the issue of land front and centre in many debates about the future of humanity, from food to energy, from climate to environment. Yet, this is not a book about the land rush; rather, it is a book about a wider issue—the politics of land. It focuses on 'rural land', not in an isolationist way, but in relation to the urban.

During the past decade, multiple global crises—involving food, fuel, energy, the environment, climate, finance, labour mobility and migration, and governance—have unfolded from and converged in a common logic: the crisis of large-scale, fossil-based, financialized industrial development. The symptoms of the crises include the fact that a billion people go hungry, although the global supply of food is more than enough to feed everyone; the continuing dependence on fossil-based energy which aggravates climate crises; the increasing role of global finance capital and indebtedness in everyday life; the rise of jobless economic growth, inequality, and mass discontent of citizens over their conditions.

State and corporate responses to these crises have largely been based on the assumption that, worldwide, there are marginal, under-utilized, empty, and available lands that can be put to efficient use to produce more food and other commodities as well as renewable energy (e.g., biofuel and hydropower). It is assumed that this land can open investment frontiers, generate employment and lead to economic development. This is hailed by neoclassical and new institutional economists as a 'quadruple win' for the state, corporations, citizens, and the environment.

It is this assumption that facilitated the recent global commodity rushes for food, feed, fuel, renewable energy, and carbon through nature conservation. All these require a supply of biomass such as soya, sugarcane, corn, palm oil, or trees. The commodity rushes required abrupt and extensive changes in land use and 'land politics'—in other words, the process of determining who gets what land and how, how much they get, for what purposes, and with what implications. In turn, this has led to what has been popularly referred to as 'global land grabbing' or the 'global land rush'.

The direct impact of these commodity rushes and emerging commodity and land frontier-making on the half of the world's population who live and work in rural areas has been far-reaching, especially in terms of recasting access to space and natural resources. The indirect spillover effects on the (peri-)urban world are known only partially. Equally important, and even less well understood, is the likelihood that the commodity and land rushes have interacted with, that is, shaped and been shaped by—the five spheres of global social life around food, climate politics, geopolitics, labour, and nation-state and citizenship.

Past commodity and land rushes, occurring across various phases of pre-neoliberal capitalism, were numerous and cyclical; they were also sectoral, often localized, intended to open new commodity and land frontiers, and extractive in character. While recent commodity and land rushes represent continuities with past cycles, there are significant differences. In our view, the recent commodity and land rushes were not simple episodic spikes in capitalist commodity markets but have caused, led to, or unleashed regime changes within commodity and land politics as well as in the five spheres of global social life.

For those who saw the solution to global crises as the 'proper' use of supposedly marginal, empty, and available land, the land rush was viewed as a rare opportunity for the world economy. In the hands of 'efficient users', such land could be exploited to double food production by 2050, promote land-based climate change mitigation and adaptation, and create investment alternatives for finance capital after the financial crisis of 2007–2008, among other things. The neoclassical economics assumption about 'marginal lands' is directed at 'monetarily low value' production systems such as shifting cultivation and pastoralism and lands that are not yet fully privatized. These defining factors shaped the geopolitics, agronomy, and political economy of the land rush. Moreover, each sectoral commodity rush ushered in its own array of actors; for example, the biofuel rush attracted those involved in energy, chemicals, auto and aviation, transport, banks, pension funds, food and agribusiness, machinery companies, and others.

But change in land use implies radical changes in land control and associated resources such as water. As a consequence, the commodity rushes have disrupted existing land politics. State actors exploited and contributed to the frenzy by manoeuvring to transform lands into a commodity, recasting rules on land politics. Numerous individual and mafia-like groupings of entrepreneurs, 'pioneers', brokers, thieves, scammers, and swindlers joined in. Initial impacts of the land rush on local communities and societies in general have been significant, albeit differentiated and uneven within and across societies. The land rush fanned the flames of inter-ethnic tension and xenophobia across the world. Multilateral institutions took note and took action. In 2011, the World Bank reported on the land grabs and recommended a 'code of conduct' for corporations.

Studying the impacts of specific land deals often entails looking at the implications for social relations around property, labour, and livelihood; this is the general method that is dominant in the current scholarship. While this is clearly important, other impacts of commodity and land rushes are frequently overlooked. If we want to understand the broader meanings and implications of commodity and land rushes, then it is vital that we examine their impacts far beyond the immediately affected communities, by inquiring into the possible structural, institutional, and political shifts in commodity and land regimes and their relation to the five spheres of social life. This is because the kind of social change that unfolds depends largely on the kind of interaction that occurs between social structures and institutions and across systems. There are no predetermined outcomes in these three-way dynamic processes. The limits and possibilities of multiple regime transitions (in commodity and land regimes and the five spheres of social life) are partly dependent on the nation-state's performance of its two permanent but contradictory tasks: facilitating capital accumulation, while maintaining a historically determined minimum level of political legitimacy (in James O'Connor's terms). Studying commodity and land rushes allows us to see and understand possible regime transformations that may be triggered not only within commodity and land politics but also in the five spheres of global social life, and how the latter, in turn, have reconfigured commodity and land regime transitions.

We build on our ongoing work on how socially just and ecologically sustainable policies can be advanced by putting together five key agendas for deep social reforms based on the principles of social justice around the *redistribution*, *recognition*, and *restitution* of wealth and power within and across societies, the *regeneration* of human–nature relations, and issues of *representation* in state–society politics. Together, these will be referred to as the 5Rs. Policies of redistribution are necessary to address the existing maldistribution of wealth and power, including land access. Policies of recognition are needed to protect the rights of peoples, especially vulnerable groups such as ethnic minorities, women, and youth, where these rights are threatened, including their right to access space, territory, and resources. Policies of restitution are needed to restore access to rights to peoples that have been stripped of them; this will include access to space, territory, land, and natural resources for those expelled from their communities, or social and health insurance for those deprived of such measures. In contrast to dominant economic perspectives, policies of regeneration combine social, political, economic, ecological, and cultural dimensions of social reforms, recognizing, for instance, that land has multidimensional meanings for people, and cannot be reduced to simple economic valorization. Policies based on the principle of regeneration aim to reintegrate narrow reform agendas, such as commodity and land regime transformations, within the broader spheres of human global social life. All the four Rs may exist only on paper if there is no sufficient social force that would fight for them and institutional mechanisms that would facilitate such political struggles in favour of the exploited and the oppressed. *Representation*, especially of the working people with diverse land questions, is key. We use the 5Rs and concepts such as land regime and land regime democratization both as an analytical lens to understand the contradictions and limits to existing mainstream land policies and politics, and as normative political reference points to see how land politics can be carried out if social justice and ecological sustainability are to be attained. A focus on the 5Rs, land regime, and land regime democratization will ultimately help us to see whether existing sectoral and fragmented approaches to land issues, policies, and politics contribute in any significant way to societal transformations towards a social justice-based and ecologically sustainable future.

Saturnino M. Borras Jr. and Jennifer C. Franco
The Hague, the Netherlands

Acknowledgements

This book is based on our extensive research on land politics in various regions of the world during the last 30 years. We have published extensively on this topic. Many of the keywords included here are concepts that we have been working on, and many of the interpretations we offer in this book are based on our own academic and activist work on land issues. For some entries in this book, we have reused, reframed, revised, updated, and repurposed previous works. For this, we acknowledge and thank the publishers of the previous publications from which we have loosely drawn elements for some entries in the current book.

In the process of developing the ideas in this book, we have accumulated so much debt of gratitude to many institutions, organizations, and individuals. They are too many to be named here, and we run the risk of missing some. Instead, we focus on a few institutions and individuals that have something to do directly with the effort to get this book project done.

For Jenny, she would like to acknowledge funding and logistical support from the Transnational Institute (TNI), especially the Agrarian and Environmental Justice Program and the Myanmar-In-Focus Program. For years of conversations, sharing, and discussion, she would like to thank Tom Kramer, Pietje Vervest, Katie Sandwell, Sylvia Kay, Carsten Pedersen, Mads Barbesgaard, Lyda Fernanda, Khu Khu Ju, Mi Kamoon, Doi Ra, Phwe Pyu, Wah Shee, Myint Zaw, and Sai Sam Kham.

For Jun, he would like to acknowledge funding support from the European Research Council (ERC) Advanced Grant (Grant Number 834006) RRUSHES-5 project, 2019–2025, as well as the Erasmus Chairs program of the Erasmus Universiteit Rotterdam (EUR). He would like to thank Veronica Goussatchenko, Chaya Raghoenath, and Eva Broer of the Projects Office of the International Institute of Social Studies (ISS) of EUR for all the management support that allowed him to focus on co-writing the content of this book.

Jun and Jenny would like to thank Ye Jingzhong and his team at the College of Humanities and Development Studies (COHD) of China Agricultural University, and the peasant movement *Katarungan* in the Philippines, especially

Danny Carranza and his team, for all the collaborative initiatives from which we have learned a lot about the politics of land.

We would like to thank Paula Bownas for the excellent copyediting, Hannah Ferguson at Routledge for encouraging us to pursue this book project and guiding us in the process, as well as Serena Watkins at Routledge for the editorial assistance.

A

Access

The 2003 article by Jesse Ribot and Nancy Peluso has altered the way we understand property and access, and the overlaps and divergences between 'bundle of rights' in property and 'bundle of powers' in their theory of access (Ribot and Peluso 2003). For Ribot and Peluso, it is important to differentiate the notion of 'access' from the concept of property and property rights (for the latter, see 'Property', this book). For them, 'access' is:

> ...the ability to benefit from things – including material objects, persons, institutions, and symbols. By focusing on ability, rather than rights as in property theory, this formulation brings attention to a wider range of social relationships that can constrain or enable people to benefit from resources without focusing on property relations alone.
>
> (Ribot and Peluso 2003: 153–154)

They argue that the main difference between property and access is found in the difference between ability and rights. For Ribot and Peluso, 'ability is akin to power' (associated with the autonomy and capacity to pursue what one thinks or wants to do), in contrast to rights, which are defined by Macpherson as 'enforceable claims' (Macpherson 1978). Ribot and Peluso then proceed to elaborate their theory of access by looking into key elements, such as 'access control', 'access maintenance', and 'gaining access'. They argue that 'access control is the ability to mediate others' access' (2003: 158), while access maintenance 'requires expending resources or powers to keep a particular sort of resource access open' (2003: 159). Gaining access is 'the more general process by which access is established' (2003: 159). They also clarify that rights-based access includes both legal and illegal access.

Finally, Ribot and Peluso explain that:

> ...property relations constitute only one set of mechanisms amongst many by which people gain, control, and maintain resource access. Our analysis expands beyond the 'bundle of rights' notion of property to

DOI: 10.4324/9781032658438-1

> a 'bundle of powers' approach to access and has advocated for locating these 'powers' within the social and political-economic contexts that shape people's abilities to benefit from resources. The categories we used to illustrate the kinds of power relations that can affect rights-based mechanisms of access were: access to technology, capital, markets, labor, knowledge, authority, identity, and social relations.
>
> (Peluso and Lund 2011: 172–173)

For the purposes of this book, it is important to understand the concepts of access and property, bundle of rights, and bundle of powers in a relational manner, especially when we focus on land access. When we look closely at a landscape and society, and try to understand land in the connected whole of production and social reproduction, it becomes clear that this is not an either/or question, that is, property rights or access; rather, layers of diverse land access permeate the mosaic of property relations, that is, private, state and commons, and the many hybrids in between.

We offer a vignette to illustrate some of the insights above. A rural household has freehold farmland and also illegally squats on land owned by the state, located in the village centre. The household allows its goats to roam around on community land, and on the village lands that are state-owned; but the grazing of the goats keeps the weeds under control. The household members gather non-timber products in the community forest for use value and exchange value. After the harvesting of paddy in the village, some members of this household go around the neighbours' paddy fields, combing through the fields for random rice stalks with grain that have been left behind; they use this to feed their chicken, which range freely on their home lot and those of their neighbours. This abbreviated mapping of a household's activities shows diverse access navigated through a mosaic of property relations.

Some everyday productive and reproductive activities could be invisible or inconsequential to outsiders, but nevertheless important to some villagers, and again, access is central. In Gambella, Ethiopia, we did field research in one of the largest land investment sites in the region, involving the expropriation of 28,000 hectares of village commons by an Indian company. Villagers were unhappy, but complied with the wishes of government partly because of the many promises made by the company and the state such as employment of the villagers in the land enterprise, social projects in the village, and so on. The company took over the allocated land, brought in bulldozers and cleared the secondary growth forest. It brought in big farm machinery, and constructed infrastructure (water pumps, etc.) fit for a large-scale industrial agribusiness plantation. Production commenced. Villagers lost access to the vast lands of the village—28,000 hectares of land that villagers had used as grazing land for their cattle and goats, for subsistence-oriented farming, as well as for gathering non-timber forest products.

The villagers felt this loss, but many remained hopeful that the company would honour its promise of providing employment. However, the jobs made available were far fewer than had been promised. But what seems to have turned the villagers decisively against the company was something else. In the village, after maize had been harvested from a farm, any maize cobs that were left behind in the field could be gathered by anyone in the village. The company produced a significant quantity of maize, but it did not allow villagers to gather any left-behind maize cobs. From the way this story was related to us, it seems that the villagers saw this as a violation of the moral economy of the village. We can express this slightly differently: that this particular type of access (access to post-harvest maize cobs), which had been established and maintained through the norms of the community, was eroded by the company.

The two vignettes above illustrate why it is relevant to navigate both property and access. We understand land access and land relations as organically embedded in society, in some ways reflecting Polanyi's argument (Polanyi 2001 [orig. 1944]). We thus see land access not in a minimalist way, limited to a farm plot or a home lot, but rather we see those farm plots or home lots as necessary parts of a broader whole. We understand access to land as something diverse, multiple, and multisited, so that an exercise to plot and connect the dots will yield a graphic image of a web (as also envisaged by Ribot and Peluso 2003), rather than just one or two lines of connection (home lot to farm plot, for example). We do not assign a vertical hierarchy of importance, for instance, asking whether access to farmland or a home lot is more important. Instead, we see a horizontal, non-hierarchical plurality of land access for production and social reproduction to produce exchange value and use value. Farmland, home lots, playgrounds for children, landing access to a riverbank to fetch water for laundry, access to a local community forest, and so on, all play a part.

Taking this view of land access, questions about the institutional aspect of property no longer involve polarizing alternatives such as private freehold or private usufruct, private freehold or state ownership, private land or commons. Instead, it becomes relevant to ask how diverse forms of access to different types of land or nature emerge within a mosaic of property relations (private, state, commons, and hybrids), and by implication, what social structures and institutions hold the mosaic together.

Boundaries are inherent to land politics, but not necessarily and always in the way that neoclassical economics would imply—that is, solid, legal, private property boundaries, often secured by material fences and protected and enforced through the coercive apparatus of the state (police, military, judiciary, and prison). If land access is inherently plural and diverse in most societies, then boundaries are equally diverse, ranging from stationary to fluid, porous to solid, long-term to transient. The stability of social relations around land access in and through the landscape, socioagroecological zone or territory is facilitated by the fluidity, malleability, and porosity of boundaries that are dynamically negotiated, contested, and transformed within these spaces. In any landscape and at any juncture, a relatively stable combination of these types of boundaries may coexist; once this pattern is challenged, with the goal of replacing it with a new combination, the political contestation could be fierce. This contestation can take different forms simultaneously, unfolding in what Kerkvliet (2009) termed 'official politics', 'advocacy politics', and 'everyday politics'.

Recommended reading

Peluso, N. L., & Lund, C. (2011). New frontiers of land control: Introduction. *Journal of Peasant Studies, 38*(4), 667–681.

Ribot, J. C., & Peluso, N. L. (2003). A theory of access. *Rural Sociology, 68*(2), 153–181.

Xu, Y. (2019). Rethinking the politics of land-use change: Insights from the rise of the industrial tree plantation sector in Southern China. *Land Use Policy, 87*, 104025.

Alliances

Land policies—both redistributive and distributive variants (see 'Redistributive reform', this book)—are about social relations from an intersectional perspective, namely, class and a range of interlocking axes of social difference: race, ethnicity, caste, gender, generation, religion, nationality. Policies that aim to redistribute ('zero sum') or distribute ('positive sum') lands potentially or actually alter land-based social relations. For example, classic land reform expropriates land from the landed classes and redistributes it to landless and near-landless working people. Distributive land policies assume that the state can take unowned and unused land and distribute it to particular social groups, as in the case of many internal colonization programmes, past and present. This assumption has increasingly been questioned and contested, especially from the perspective of Indigenous peoples whose lands were often the ones taken by the state for its colonization programmes.

The politics of land can be best understood through the alliances that emerge for and against particular types of land policies. These contending alliances are not always a straightforward story of the political and economic ruling classes versus the traditionally exploited and oppressed working classes. The class configuration in such alliances is far more complex than simplistic assumptions about land and power might suggest.

In conventional agrarian studies, land often means agricultural land. Land redistribution means taking land from large private estates and redistributing it to landless and near-landless working people. At best, the official goal is to create a smallholder-based agricultural sector; at a minimum, the aim is to end the socioeconomic and political power of the traditional landed class. There are conventional oppositional forces to this kind of land policy, most prominently the landed classes whose main income is derived from collecting land rent from those who use their lands (tenants and sharecroppers, for example). This privilege also brings political power that can extend to landed classes' ability to occupy government positions. When they are also entrenched within the state and its coercive apparatus, the power of the landed classes to block reforms is enormous. Thus, police, military, and judiciary often come to serve the interests of the landed classes. Rich farmers—those who have relatively big farms, but who still work the land directly and usually employ workers and use modern technology (farm machines, etc.)—often take the side of the landed classes politically and ideologically. They do not want their lands to be affected by redistributive land policies, and their aspiration is to continue to expand the area of their accumulated land.

It is not surprising that other sections of the dominant socioeconomic classes ally with the landed classes, politically and ideologically. Families are often connected through marriages or interlocking corporate investments. However, there have also been occasions in history when non-landed economic and political elites chose to manoeuvre against the landed classes for various reasons, including economic, political, and ideological. Some industrialists and finance capitalists, for example, see the exploitative and oppressive feudal or semi-feudal agrarian system as limiting their own class interest in expanding the base of their market or sources of cheap labour. This section of the ruling classes has at times supported liberal, bourgeoise forms of land redistribution. In fact, many liberal land reforms were able to succeed because the ranks of the ruling classes were divided, weakening opposition to reforms to the point that it could not block democratic challenge. Pro-reform factions of ruling elites have sometimes managed to entrench themselves within the state, enabling them to mobilize the coercive apparatus of the state to advance redistributive land policies.

Strategic allies of landless and near-landless working people are usually to be found within other sections of the working class, both rural and urban, as well as political groups (political parties, electoral parties, other social movements, especially feminist movements). In some instances, elements of influential religious institutions may also be part of this alliance, for political and ideological reasons. All the relatively successful redistributive land policies to date had behind them broad-based alliances among these societal forces, along with a section of the ruling class and political elite.

The description above reflects the dominant combination of forces for and against redistributive land policies that dynamized struggles for land redistribution during much of the twentieth century. However, there is a caveat that needs to be pointed out. Indigenous communities worldwide have always been wary of redistributive and distributive land policies because there have been instances when it was their lands that were taken by the state for distribution, either through a land reform programme or an internal colonization campaign. This tension between Indigenous peoples and peasant communities remains quite palpable and relatively widespread in the contemporary period.

The late twentieth century and early twenty-first century have been marked by the rise of urban sprawl and climate change politics. The politics of land, including land struggles for redistribution, have become complicated as old and new types of alliances make and remake one another; potential and actual political forces against redistributive land policies that favour working people have become more diverse, stronger, and more complex to tackle. The two most powerful political forces against redistributive land policies today are real estate conglomerates and big conservation organizations. Real estate companies have been at the forefront of grabbing farmlands and converting them to high-value residential and commercial enclaves. Meanwhile, neoliberal conservation organizations continue to push to expand the geographic areas under their control at the cost of shrinking the land area for productive and social reproductive activities of rural working people (Brockington and Duffy 2011).

While most farmland struggles today are carried out by peasant movements and Indigenous communities and their allies, especially NGOs, there are other actors whose full potential as allies in land struggles is yet to be realized. One such set of actors is environmental and climate justice movements, as many current environmental and climate justice issues are connected with land (Martinez-Alier et al. 2016; He 2023). Another is the disproportionately large section of the working class that exists in the informal sector as semi-proletariat or precariat, many of them in urban and peri-urban spaces. They have a lot to gain from 'land regime democratization' (this book) (Davis 2006). Whether such broad alliances are possible to build, and if so, how, remains to be seen. But if struggles for land are to be embedded within the logic of the current conjuncture marked by joblessness, precarity, and climate crisis, then such a coalition seems inevitable.

In short, alliances among state and social forces—for and against redistributive land policies that aspire to benefit working people—have become even more complex since the late twentieth century, in the context of massive urbanization and the rise of climate change politics. Nevertheless, class politics remains central to our understanding of the fundamental interrelationship of state and social forces, even though these are further shaped by interlocking relations around race, ethnicity, caste, gender, generation, religion, and nationality from one society to another (see also 'Sandwich strategy', this book).

C

Class

Class is "the social relation of production between classes of producers (labour) and non-producers" (Bernstein 2010: 124). This political economy definition of class is fundamentally located in a social group's relation to the means of production, by virtue of which individuals associated with such a group tend to do similar undertaking in order to survive or accumulate, and thereby relate with other social groups in the process. For example, a group of people who live in the countryside who do not have their own land but needed to farm may end up using the land of an owner of a big estate who does not want to work the land—in exchange of a rent, whether in cash or in terms of crop share. This is a tenant farmer-landlord class relation. Further, the share–tenants may end up borrowing money for subsistence from individuals who have money—in exchange of interest, again, either in cash or in crop share. This is a tenant farmer-moneylender class relation. This kind of social relation entails that a "'surplus' has to be produced above the needs of the producers for their consumption, replacement and ceremonial funds, in order to support dominant classes of non-producers" (Bernstein 2010: 21). Bernstein concluded: "The capacity to *appropriate surplus labour*—labour beyond what producers expend on their own reproduction—signals social relations of *exploitation* (Bernstein 2010: 21, original emphasis).

In the study of the politics of land, without class analysis, our inquiry does not go anywhere because we will not even know about the pattern in which some individuals gain benefits, while others lose. Without class analysis, we either end up with random individuals (such as the neoclassical economics notion of 'rational individual'), or undifferentiated social groups like 'community', or sectoral categories like 'rural people', 'urban people' or 'people of the land' that conceal more than reveal dynamics of social relations that are the causes and consequences of land-based social change. Yet, with class analysis alone, our analysis does not go as far as necessary. This is because "class relations are *universal but not exclusive* 'determinations' of social practices in capitalism", as Bernstein pointed out

DOI: 10.4324/9781032658438-2

(2010: 115, original emphasis). He further elaborated that "they intersect and combine with other social differences and divisions, of which gender is the most widespread and which can also include oppressive and exclusionary relations of race and ethnicity, religion and caste" (2010: 115). Class and interlocking social relations are thus central to our understanding of land politics and have been a guiding reference point for us in writing the entries in this book.

Further elaboration is warranted.[1] With the centrality of class relations, class politics become fundamental. How class relations emerge and how they are transformed into a political force—the transformation of 'class-in-itself' to 'class-for-itself'—are two inseparable social processes, politically and analytically. And class is co-constituted with other axes of social difference—race, ethnicity, caste, gender, generation, religion, or nationality—our understanding of 'class agency' and class-for-itself is necessarily grounded on these mutually constitutive relations. This brings us to an important analytical signpost about class and class consciousness advanced by E.P. Thompson, orbiting around his main point that class is relational and historical, that is, the notion of historical relationship.

Thompson argued that "the relationship must always be embodied in real people and in a real context… we cannot have two distinct classes, each with an independent being, and then bring them *into* relationship with each other" (Thompson 1991 [orig. 1963]: 8, original emphasis). He continued:

> class happens when some men, as a result of common experiences (inherited or shared), feel and articulate the identity of their interests as between themselves, and as against other men whose interests are different from (and usually opposed to) theirs.
>
> (Thompson 1991 [orig. 1963]: 8–9)

He emphasized that:

> The class experience is largely determined by the productive relations into which men are born – or enter involuntarily. Class consciousness is the way in which these experiences are handled in cultural terms: embodied in traditions, value-systems, ideas and institutional forms. If the experience appears as determined, class consciousness does not. We can see a *logic* in the responses of similar occupational groups undergoing similar experiences, but we cannot predict any *law*. Consciousness of class arises in the same way in different times and places, but never in just the same way.
>
> (Thompson 1991 [orig. 1963]: 9, original emphases)

Thompson stressed the historical dimension of class: "these are historical questions. If we stop history at a given point, then there are no classes but simply a multitude of individuals with a multitude of experiences" (Thompson 1991 [orig. 1963]: 10). But he elaborated:

> if we watch these men over an adequate period of social change, we observe patterns in their relationships, their ideas, and their institutions. Class is defined by men as they live their own history, and, in the end, this is only its definition.
>
> (Thompson 1991 [orig. 1963])

He concluded that "we cannot understand class unless we see it as a social and cultural formation, arising from processes which can only be studied as they work themselves out over a considerable historical period" (Thompson 1991 [orig. 1963]: 10–11).

Recommended reading

Bernstein, H. (2010). *Class dynamics of agrarian change*. Halifax: Fernwood Publishing. For Open Access ebook edition: https://practicalactionpublishing.com/book/2571/class-dynamics-of-agrarian-change

Climate change politics

The politics of climate change is broad. In this entry, our treatment of this concept is quite narrow, focusing on the relationship between climate change politics and the politics of land. Climate change politics is inherently linked to land and land politics; it is a phenomenon that separates the politics of land in the twenty-first century from its counterpart in previous centuries. Climate change is not new: it started as soon as capitalism gained momentum globally (Malm 2016; Moore 2017). But climate change *politics* is relatively new, emerging in the late twentieth century. A clear understanding of the politics of land and its historical role in the development of global capitalism will help us understand the causes of climate change, and a better understanding of climate change will help us understand the historical trajectory of land politics. The current conjuncture is marked by the inherent links between climate change and land: resolving the challenges of one depends on resolving the issues of the other. How land and climate change are connected is discussed in many ways from different perspectives (e.g., Foster 1999; Holleman 2018; IPCC 2019; Saito 2023). The brief discussion here is specifically focused on the politics of climate change mitigation and adaptation narratives and how these have become central to thinking about the politics of land in current times.

We define climate change politics as the dynamics operating in the spheres of social structures, institutions and politics, namely, social relations; policies, treaties, laws, procedures, norms; projects, programmes, narratives, ideas; advocacies, mobilizations, and social movements; memories, rumours, or gossip – separately and collectively, among and between different social classes and groups within the state and in society, that set and shape the meanings of climate change, its causes and consequences, how it can be addressed, by whom, where and when.

Climate change and land are intimately linked. The commodification of nature has resulted in or exacerbated the climate crisis. The Industrial Revolutions of past centuries introduced the extensive use of fossil fuels and methods of production of commodities and patterns of circulation, consumption and waste that brought us to the current climate crisis. A subset of these industrial revolutions was the rise of large-scale industrial agricultural production that ruptured the 'metabolic interaction' between humans and the earth, which 'prevents the return to the soil of its constituent elements consumed by man in the form of food and clothing; hence, it hinders the operation of the eternal natural condition for the lasting fertility of the soil' (Marx 1976a: 637–638). Therefore, this type of production 'only develops the techniques and the degree of combination of the social process of production by simultaneously undermining the original sources of all wealth—the soil and the worker' (Marx 1976a: 638).

One outcome of this production system is the rise of monoculture plantations (León Araya 2023). Even during its early period, the system struggled to replenish soil fertility eroded by the industrial production method and the long distances between sites of production, consumption, and waste. Profit-making was maintained through the continuous expansion of forest clearing and the mining of natural soil fertility in settler economies, combining it with the use of animal bone and guano as fertilizer to sustain profit-making through large-scale production. This was later replaced by mined minerals (phosphate, potash) and synthetic fertilizer (nitrogen) tied to the fossil complex. The level of agricultural production achieved in this system cannot be sustained indefinitely, as the supply of minerals and fossils is finite. As long as this system is seen as key to the broader economic growth framework that subordinates nature-based solutions (NBS), those NBS are unlikely to make any significant difference. Producing commodities to generate profit epitomizes the fundamental contradiction between capitalist production and ecology because it results in the non- or under-valuation of nature, the opportunistic use of nature to temporarily fix some of the crises in capitalism and the emergence of a global chain of destructive production–consumption systems (O'Connor 1998; Gerber 2025). It is a global chain of destruction and pollution.

The politics of land plays a central role in the reproduction of this global chain of destruction. That land will, or should, play a key role in addressing the climate crisis is a popular view, as the concept of NBS shows. Nevertheless, how this is being framed is fiercely contested. In the neoclassical economics approach, a key starting point to the Anthropocene challenge with regard to climate change is the assumption that particular types of production and reproduction systems are 'economically inefficient and ecologically destructive'. Specific targets include mobile small-scale pastoralists and swidden agriculturalists, as well as forest dwellers and wetlands users. In many cases, these are Indigenous peoples and other marginalized ethnic minorities. Based on this assumption, mainstream neoclassical views take a two-pronged approach. One is to seize resource control from these land users and reallocate the resources to 'economically efficient' users who can observe international standards around ecological sustainability.

The other is to monetarily compensate these land users in exchange for their commitment not to use or access their land the way they used to do, such as in various forms of fortress conservation or carbon sequestration projects involving forests. The assumption that these production systems are economically inefficient and ecologically destructive is highly questionable but has been extremely influential. It forms the foundation of the NBS and 'net zero emissions' strategy which is the main thrust of efforts to address the climate crisis. Initiatives that are deemed to be economically efficient and ecologically sound can then be rolled out. Existing production systems that are considered inferior are replaced by wind and solar farms; lands are submerged amid hydropower megaprojects, destroyed or polluted by rare earths mining.

Our definition of climate change politics allows us to separate, but at the same time connect together, politics and policies or projects—categories that are often muddled in the literature. Formally constituted and recognized climate change policies or projects are one of the components of climate change politics—the same relationship that land policies have to land politics. Labelled as climate change mitigation or adaptation measures, such policies and projects include 'Bio-energy with Carbon Capture and Storage' (BECCS) as well as officially recognized big forest conservation projects and REDD+ (Reducing Emissions from Deforestation and Forest Degradation in Developing countries) (Corbera 2012). But as many scholars working from political economy and political ecology perspectives (e.g., Paprocki 2019) remind us, policies and projects are not the only things that matter. Social relations of production and ecological and political conditions of nature (land, water, forest) could be significantly altered not only by climate change (e.g. droughts, floods) or formal climate change policies or projects but also by wild speculations, rumours, gossip, spectacle or frenzy inspired or triggered by measures purported to address climate change.

Commodification of, and therefore profiting from (ideas about), responses to climate change includes spectacular projections or anticipation of forthcoming disaster and ruin, regardless of whether such projections have a material basis (Paprocki 2022). For example, many land brokers and speculators who have orchestrated the grabbing of land from ordinary villagers in different societies during the past decade have done so by invoking the urgent need to address climate change through large-scale investments, such as establishing a plantation to cultivate feedstock for biofuel production—whether biofuel production actually takes place, is seriously planned, or is only wildly speculated (Fairhead et al. 2012; Hunsberger et al. 2017; Franco and Borras 2019). This is how most green grabbing has been operationalized (Fairhead et al. 2012).

Looking at climate change *politics* means subjecting climate change mitigation and adaptation narratives and practices to a reality check. It allows for a politicized and historicized analysis of social processes, the evolution of climate change mitigation and adaptation and social class relations and a scrutiny of terminologies such as 'vulnerability' and 'scarcity' (following Ribot 2014; Scoones et al. 2019). In our definition, climate change politics can be crudely categorized as progressive or regressive. By progressive climate change politics, we mean 'those social dynamics that shape the meanings of climate change, its causes and consequences, how it can be addressed, by whom, where, and when in ways that advance "agrarian justice" and "climate justice", separately and together' (Borras et al. 2020a: 2). By regressive climate change politics, we refer to 'social and political processes that block, undermine or reverse advocacy for or gains in progressive climate change politics' (Borras et al. 2020a: 2).

Recommended reading

León Araya, A. (2023). Monocrops. *The Journal of Peasant Studies, 50*(3), 797–808.
Paprocki, K. (2022). Anticipatory ruination. *The Journal of Peasant Studies, 49*(7), 1399–1408.

Climate of global land politics

The climate of global land politics during the first three quarters of the twentieth century was hot on expropriation and redistribution—classic redistributive land reform—whether with a capitalist or socialist perspective, pursued through armed revolutions or through liberal democratic political processes (Wolf 1969).[2] The politics of land at that time was tightly linked to debates around national development, especially through urban development and industrialization, and the role of agriculture and the rural sector in this process (Akram-Lodhi et al. 2007; Spoor 2008; Kay 2009).

The climate of global land politics in the twenty-first century is very different. It is not that the politics of land is not a popular agenda; it is – but for quite different reasons and through different measures. In the era of global urbanization and climate change, land politics has continued to occupy a prominent centre stage in development policy debates, including around conservation, climate change mitigation and adaptation, urban landscape and so on. What then is the general political position on land in the context of classic agrarian questions, that is, in terms of democratizing land control for broad-based and equitable development? The short answer is that the overall political picture is bleak.

When talking about land and the politics of land, the general tendency is both too local and too sectoral, for example, land policy for specific local communities, or land policy for a particular sector, such as farmland or grazing land. These are important and help ground one's perspective on the politics of land. However, this approach has its limits: it tends to be blind to the bigger picture, broader systemic political narratives, state and social forces and dynamics of global capitalism that are a key driving force in changes in the politics of land. A better approach is one that consistently combines the local with the global, and vice versa, the parts with the whole (discussed in the concept of 'Land regime', this book). One of the important themes in the politics of land is the climate of land politics on a global scale. While we do not want to succumb to the determinism of global forces in shaping local dynamics, neither do we subscribe to too localist a perspective. Bearing this caveat in mind, we argue that the current international climate of land politics is characterized by five interlinked political acts which we summarize as: *roll back*, *contain*, *block*, *reinterpret*, and *promote*. The neoliberal ideological position gained early ground in land policies in the 1970s, as in the land reform reversal carried out by the Pinochet regime. Since then, land has become a central object of neoliberalism: the goal is the commodification of land and nature and a key mechanism to achieve this is the promotion of individual private property rights in land. Thus, these five mutually reinforcing political acts, emanating from a single social force (the commodification process) and neoliberal ideology, have converged to construct and maintain the current structure of the global land politics architecture. These are wilful, orchestrated actions by social groups within the state and in society, in pursuit of particular socioeconomic and political interests.

In places where there have been gains in redistributive land policies under varying regimes, either socialist or capitalist, the thrust has been to *roll back* those gains. The rollback campaign started at the onset of global neoliberalism, beginning with the partial rollback of land reforms initiated by Frei and Allende in Chile, immediately after General Pinochet seized state power through a military coup (Kay 2002: 470–471). Since then, this rollback campaign has gained momentum. The dismantling of regulatory institutions in land, such as state cooperatives and collective farms, spread rapidly—if unevenly—in the 1990s (Ho and Spoor 2006; Lahiff et al. 2007; Spoor 2008). Institutions that regulate land sales and rental markets such as land size ceiling laws, or those that actively promote community or collective arrangements, such as the *ejido* system in Mexico to prevent land concentration, have increasingly been deregulated (Zoomers and van der Haar 2000; Akram-Lodhi et al. 2007). State support to beneficiaries of land redistribution has been withdrawn.

In situations where there were ongoing redistributive land policies, the overall mainstream strategy has been to *contain* the scope, pace, and momentum of such policies. Only a handful of countries have been able to pursue progressive redistributive land policies during neoliberalism, namely, Brazil and the Philippines since 1988 (Wolford 2010; Franco 2011); South Africa, with different variants of a hybrid market-led/state-driven land redistribution

(Ntsebeza and Hall 2007); Colombia; Zimbabwe since 2000 (Moyo 2011; Scoones et al. 2011); Mozambique through its 1997 Land Law and Indonesia (Lund 2023) with the retention of its Basic Agrarian Law (BAL) of 1960. In each of these cases, the strategy deployed by mainstream political forces was to limit and contain. Outright abandonment of existing redistributive land policies may be too politically risky because of possible popular protest. The strategy for containing such policies involves slowing down the speed of implementation, limiting the scope, defunding the land agency or ministry, or rendering a potentially redistributive land policy dormant—as seen, in varying contexts and extents, in the countries referenced in this paragraph.

In settings with significant numbers of landless or land poor, where land redistribution and restitution are needed but no appropriate policies or laws exist, the mainstream strategy deployed has been to *block* any land policy that is redistributive or restitutive in character. A classic example is Myanmar where the World Bank, USAID, and other international agencies that were advising the government before the February 2021 military coup pushed for market-oriented land policies, essentially blocking the possibility of system-wide redistributive and restitutive land policies (Ra and Ju 2021; Ra et al. 2021). Arguably, the most pervasive form of the blocking strategy is having a 'no policy' policy. This refers to societies where there is a compelling case for redistributive land policies, but where governments choose not to adopt any policies to achieve this end. The Indonesian BAL of 1960 has been retained but kept dormant. Even those laws and policies that sought to regulate the frenzied land rush of the past decade have been applied only peripherally; regulation has been used to facilitate large-scale land deals by minimizing collateral damage such as expulsions of affected villagers without monetary compensation.

In settings where rollback, containment or blocking are not politically feasible, an alternative strategy is to use and coopt the generic labels that are politically safe and broadly acceptable at a given conjuncture, but then *re-interpret* what they actually mean, turning them in the direction of market-oriented processes. The easy targets here are terms that are broadly popular, with meanings that can be manipulated, such as 'land rights' and 'land tenure security'. Land rights become private individual property rights, and land tenure security becomes formalization and ratification of what exists, devoid of any redistributive spirit. Land reform is interpreted to be a buy-and-sell real estate transaction through voluntary market-assisted land reform.

These four political acts, separately and together, ultimately seek to *promote* market-friendly land policies. While neoliberal land policies are constituted, interpreted, and implemented in a highly uneven manner, idealized versions take on the defining features of privatized, deregulated and demand-driven land policies aimed at stimulating land sales and rental markets, joint ventures, and credit markets. This can be facilitated by ensuring the availability of transparent data and information about private property in land that can in turn allow for informed transactions—by willing sellers and willing buyers—which require minimum bureaucracy and are enforceable with minimal transaction costs. The objective is to use land and nature in the most economically efficient manner (Pereira 2021). The notions of appropriate land use and appropriate land users are defined by a mainstream concept of economic efficiency, centred around the maximization of scarce land resources for a free market-driven economic growth that could satisfy the profit- and utility-maximizing impulses of individuals (Deininger and Byerlee 2011).

Formal private land titles can ensure the 'security' necessary to facilitate dynamic land and rental markets, joint ventures and credit markets. In this context, security means, primarily, security for owners of capital – from corporate agribusiness to finance capital, from traders to rich farmers (Deininger 2011). It is less about giving working poor people security from discrimination, or from being coerced to give up control of land. This is part of the dominant neoliberal narrative in which 'investment' becomes synonymous with 'corporate investment', consciously or unconsciously dismissing the two other equally important (if not more important) forms of investment, namely, public investments and investments of working poor people themselves (Kay 2014). The guiding principle is to ensure that the most economically efficient users (in neoclassical economics/new institutional economics terms) are able to access and use land in the most economically efficient manner—that is, realizing the greatest profit that can be derived from economic undertakings. This might entail individual or group rights, freehold title or formalized customary land tenure, or a combination of these; the bottom line is whether it will allow for land sales and rental markets, joint ventures and credit markets, to emerge and flourish, thereby assuring would-be capitalist investors of the security of their investments.

The political acts of rollback, containment, blocking, re-interpreting and promoting have resulted in the current architecture of global land politics. As such, the land policy structure is an inherent component of global neoliberal capitalism where it serves at least two major roles: as a strategy to address the capitalist crisis of over-accumulation (Harvey 2003) by using land/nature for cheap inputs in production; and, at the same time, as a strategy to address the ecological crisis of capitalist production by expanding the role of land/nature as sink for production waste (O'Connor 1998; Fraser 2021; Martinez-Alier 2021). This is not an original political strategy. The roll back, contain, block and re-interpret strategies against redistributive land policies that emerged out of socialist or anti-colonial revolutions or electoral victories of left-wing political parties were deployed by conservative political forces worldwide during the Cold War, from the 1950s to the beginning of neoliberalism in the early 1980s. These forces included political coalitions like the 1960s' US-driven Alliance for Progress as well as key individuals such as Roy Prosterman and Wolf Ladejinski (Putzel 1992). During the Cold War, the roll back–contain–block–reinterpret playbook fed into the strategy to promote counter and/or pre-emptive land policy reforms. These took the form of formal land titling, land resettlement or internal colonization, limited liberal redistributive land reforms, or a combination of these. These types of intervention were described by Diskin (1989), in the context of El Salvador, as reforms that prevent change. The nature of the contemporary roll back–contain–block–reinterpret–promote political acts is the same as that observed by Diskin decades ago.

Recommended reading

Franco, J. C., & Borras Jr, S. M. (2021). The global climate of land politics. *Globalizations, 18*(7), 1277–1297.

Climate-smart land politics

In this book, we define climate-smart land politics as neoliberal land politics sensitized to the recent corporate reinvention of agriculture and discovery of market-based climate change mitigation and adaptation initiatives. This concept allows us to make sense of trends in narratives on what kinds of land politics to pursue, given climate change politics, and how. We would argue that neoliberal land politics has been mainstreamed much earlier than climate change politics and that the rise of the latter has been increasing tapped to reinforce the former. These dynamics can be seen by looking at Climate Smart Agriculture (CSA), the neoliberal framework on agriculture embedded in the mainstream narrative around climate change.

Officially, CSA claims to have three aims: increased productivity, increased resilience, and reduced emissions. This is partly based on the projection that the world population will reach 10 billion by 2050, and on the related assumption that global food production would need to be increased by 50% from its current level. Central to CSA is a particular kind of land politics. In the 2016 World Bank Group Climate Change Action Plan (World Bank 2016), faith in market-based mechanisms and associated approaches with respect to natural resource (re)allocation and use is reaffirmed in an important section entitled, 'Climate-Smart Land Use, Water and Food Security'. The World Bank document is silent on shifting cultivation, a form of agroforestry that is widely practised globally and that, after a long period of being maligned and criminalized, is slowly (re)gaining support from some climate justice advocates in light of its many contributions to climate change mitigation and adaptation. Its silence on this issue does not mean that the World Bank supports shifting cultivation; rather, this silence, alongside the emphasis on the CSA triad of productivity, resilience, and reduced emissions, resoundingly suggests that shifting cultivation is on the list of users and uses that the World Bank thinks will have to go. This list includes peasant and Indigenous peoples' production systems, shifting cultivation, and small-scale pastoralism, while industrial farming systems and technology are promoted instead—genetically modified organisms, modern seeds, agribusiness plantations, big conservation initiatives, capital-intensive feedlot livestock ranching, and so on. The World Bank document is also silent on how and how much an expanded industrial agriculture sector contributes to greenhouse gas (GHG) emissions. More importantly, it has nothing to say on land redistribution, recognition, and restitution, and how these measures may have relevance for climate change mitigation and adaptation questions. Neither does the FAO's CSA Sourcebook (based on the massive 570-page Sourcebook of 2013) (FAO 2013) mention these three key social justice land policies, which is especially ironic in light of the FAO's key role in developing and implementing the 2012 Voluntary Guidelines on the Responsible Governance of Tenure of Land, Fisheries and Forests in the Context of National Food Security (VGGT) (Franco and Monsalve 2018). These guidelines recognize all three of these measures as part of a new international standard on land, fisheries, and forest tenure regulation. Ignoring the social justice aspects of the guidelines, the FAO CSA Sourcebook adopts or follows the World Bank's narrow 'tenure insecurity' language and framework to explain the lack of investments in land, whether from farmers themselves or corporate entities and leaves open what kind of resource tenure regime could guarantee 'security' for marginalized land users and holders. What the World Bank terms climate-smart land use is essentially its answer to who ought to get what land, for how long, and for what purpose, in light of the three neoliberal aspirations of productivity, resilience, and reduced emissions.

The contemporary processes of the global land rush and the natural resource-oriented climate change mitigation/adaptation narrative are recasting the political economy of land, water, fisheries, and forests in the rural world and reconfiguring how capital penetrates agriculture and the countryside. Ostensibly, the twin objectives are to combat what is assumed to be the inefficient and destructive use and users of scarce natural resources, mostly in the countryside of the Global South. Agriculture is at the intersection of these purported twin aims. Three broadly distinct dimensions, reformulated as questions, can be distilled out of this narrative in relation to agriculture and rural livelihoods: (1) what kinds of agriculture and rural livelihoods are 'efficient and productive' enough to sustain capital accumulation imperatives? (2) what kinds of agriculture and rural livelihoods are 'ecologically sustainable and less precarious' for the population? and (3) where do efficient and sustainable agriculture and rural livelihoods end and a broad range of nature conservation/commodification begin? The last of these questions also trains a spotlight on the concept of 'sustainable development' and on the validity and relevance of pursuing Sustainable Development Goals (SDGs) as part of this architecture. Meanwhile, these three dimensions are addressed through a variety of mutually reinforcing CSA and neoliberal nature conservation policies and initiatives.

Crucially, the lynchpin that links the three dimensions, discursively and materially, is land. The common denominator in political terms between CSA and nature conservation and commodification is land control. A common policy obsession for these overlapping market-oriented advocacies is land policy. Whether one is pushing for an exponential increase in food production or in carbon offsetting, one necessarily has to confront the issues of land, land control and land policy. A critical examination of this mainstream narrative shows an ideological stand common to all: the narrative is intrinsically averse to social justice-oriented land redistribution, recognition and restitution. It seems to us that what the combined processes suggest is a kind of predatory land politics that serves the neoliberal campaign against what it assumes and perceives to be the inefficient and destructive use of scarce natural resources. This is climate-smart land politics.

Climate-smart land politics is necessary to allow for responses that can truly help expand the frontier of capital accumulation in the midst of climate crisis, while at the same time making sure that such responses are themselves at the frontier of profit-making. This is the logic behind the mainstream, popular framework for climate change mitigation and adaptation, namely, NBS such as carbon sequestration conservation projects. But even if we assume that projects within the NBS framework do work, at best, this approach amounts to scattered and incremental small reforms that are not capable of effectively addressing climate change as a systemic, planetary crisis. Most of the NBS projects are land-plot or block oriented, in which a demarcated space has to be declared an official project before it can be considered a proper mitigation and adaptation measure. There are various flaws in this approach. Blocks of land not officially considered as mitigation and adaptation projects, such as regular community forests, are not necessarily protected always and everywhere. Demarcated spaces for forest conservation may not conserve carbon (as many audits have shown), and even if they do, they may trigger spillover effects to adjacent places. For example, local villagers whose livelihoods are curtailed by the conservation project may migrate to nearby or distant places outside the official conservation area and start new clearing or extraction activities in order to survive.

Moreover, carbon savings in an officially demarcated area may easily be cancelled out by, for example, forest clear-cutting to open the way for a new soya plantation in another part of the country. A block of peatland excluded from oil palm plantation expansion can be cancelled out by a block of wetlands converted into an industrial fishpond operation. This approach often results in perverted intervention. For instance, natural forest is destroyed to be replaced with an industrial tree plantation of fast-growing trees; this is declared a mitigation and adaptation project, with profits generated and collected. The production and consumption of biofuels have steadily increased over time, and while biofuels may not contribute to carbon emissions on the consumption side, production processes (forest clearing, plantation maintenance, industrial-chemical production systems, long-distance transport of biofuel commodities) can cancel out many times over whatever the consumption-side gains might be.

One of the problems with this approach is its piecemeal take on land, that is, its focus on the level of plots or blocks, at scales well below a land regime. Ultimately, however, the main flaw of the NBS approach, the currently dominant strategy to address climate change, is that it does not truly aim to change the logic that underpins the climate crisis: the generalized commodification of land and life in which profit-making is the engine that keeps everyday life running, even while it destroys the ecological basis of that life. Hence, at best, NBS can generate small and scattered oases of positive gains in carbon offsets and savings within an ever-expanding wilderness of death and destruction. It is like squeezing a balloon: squeeze it here, it pops up in another place. Different generations and categories of climate change mitigation and adaptation strategies are significantly land-based (biofuels, REDD+, net zero emissions, NBS). Separately and collectively, these have not made any significant impact in reversing the level of climate crisis. We would argue that this is because they fail to address the global logic of destruction as the root cause of the climate crisis.

Recommended reading

Clapp, J., Newell, P., & Brent, Z. W. (2018). The global political economy of climate change, agriculture and food systems. *The Journal of Peasant Studies, 45*(1), 80–88.

Taylor, M. (2018). Climate-smart agriculture: What is it good for?. *The Journal of Peasant Studies, 45*(1), 89–107.

Commoning

The contemporary broad left has picked up on the concept of commons in relation to struggles within and against capitalism, and has emphasized three things. First, the cycles of crisis inherent within capitalism result in a crisis of social reproduction; capitalism is forced to rely on commons in order to address this crisis of social reproduction. There are echoes here of Karl Polanyi's notion of double movement, that is, that society will mobilize to take care of confronting the crisis (Polanyi 2001 [orig. 1944]). Second, capitalism will at the same time continue to try to enclose the remaining commons, as exemplified by the global land grabbing of the twenty-first century, to acquire raw materials free of cost or at very low cost to fulfil its quest for a never-ending process of economic growth and profit making (Harvey 2003). Third, struggles within and against capitalism increasingly involve the process of protecting, consolidating or constructing commons in the broadest sense—from the traditional meaning of the term as land commons, to more recent notions such as 'digital commons'—as alternatives within and outside of capitalism. The meaning and usage of the term commons in this political context is far looser than the narrower definition that we used earlier (see 'Access', this book) and is more directly related to creating different forms of collectivities in relation to the struggles within and against capitalism. The strategy for protecting, consolidating, expanding or creating and building commons is generally referred to as 'commoning', with a focus on political practice and contingency, agency, and community (De Angelis 2013).

Commons are not things, they are social relations; they are not fixed givens; they are a process and are created. The term 'commoning' highlights social relations and political process; it is the political process of (re)building commons. But this does not mean that material things are not important. Caffentzis and Federici (2014: 101) argue that 'commons must guarantee the reproduction of our lives... Water systems, lands, forests ... are indispensable to our survival. Here too what counts is the collective nature of the reproductive work and the means of reproduction involved'.

Moreover, the emphasis is on a particular type of commons, namely, 'anti-capitalist commons'. Caffentzis and Federici (2014: 100) clarify that the commons they refer to as being urgently needed 'aim to transform our social relations and create an alternative to capitalism'. They elaborate:

> Anti-capitalist commons, then, should be conceived as both autonomous spaces from which to reclaim control over the conditions of our reproduction, and as bases from which to counter the processes of enclosure and increasingly disentangle our lives from the market and the state. Thus they differ from those advocated by the Ostrom School, where commons are imagined in a relation of coexistence with the public and with the private. Ideally, they embody the vision that Marxists and anarchists have aspired to but failed to realize: that of a society made of 'free associations of producers', self- governed and organized to ensure not an abstract equality but the satisfaction of people's needs and desires. Today we see only fragments of this world... but already the commons we build should enable us to gain more power with regard to capital and the state and embryonically prefigure a new mode of production, no longer built on a competitive principle, but on the principle of collective solidarity.
>
> (2014: 101)

In terms of land politics, advocates of anti-capitalist commons are logically supportive of the idea of building various forms of common property regimes in various types of resources (public, private, common, club; see separate discussion on 'Commons', this book). Indigenous peoples' territories; various forms of communal, community, cooperative and collective ways of governing access and use of peasant lands and community-based forest governance have all gained in popularity among social activists since the beginning of the twenty-first century.

How exactly do we protect, consolidate, expand, and create anti-capitalist commons? In other words, how do we pursue commoning? Caffentzis and Federici have outlined a practical call to action which offers intellectual and political reference points, a set of propositions for discussion, rather than a prescription (see Box 1).

Box 1. Some practical anti-capitalist commoning reference points

'To guarantee our reproduction "commons" must involve a "common wealth", in the form of shared natural or social resources: lands, forests, waters, urban spaces, systems of knowledge and communication, all to be used for non-commercial purposes.'

'One of the challenges we face today is connecting the struggle over the public with those for the construction of the common, so that they can reinforce each other. The public is a state institution that assumes the existence of a sphere of private economic and social relations we cannot control. What we call "the public" is actually wealth that we have produced and we must re-appropriate it. It is important to maintain the distinction between public and common.'

'Commons require a community. A specific community is created in the production of the relations by which a specific common is brought into existence and sustained.'

'Commons require regulations stipulating how the wealth we share is to be used and cared for, the governing principles being equal access, reciprocity between what is given and what is taken, collective decision making, and power from the ground up, derived from tested abilities and continually shifting through different subjects depending on the tasks to be performed.'

'Equal access to the means of (re)production and egalitarian decision making must be the foundation of the commons.'

Caffentzis and Federici (2014: 101–102)

Caffentzis and Federici (2014: 103) conclude that 'anti-capitalist commons are not the end point of a struggle to construct a non-capitalist world, but its means. For no struggle will succeed in changing the world if we do not organize our reproduction in a communal way...'. Exactly how the idea of anti-capitalist commons and commoning resonates with various existing land struggles and land movements is a theoretical and empirical question that needs to be explored more carefully in the context of the twenty first century.

Recommended reading

Caffentzis, G., & Federici, S. (2014). Commons against and beyond capitalism. *Community Development Journal*, *49*(suppl_1), 92–105.

De Angelis, M. (2013). Does capital need a commons fix?. *Ephemera*, *13*(3), 603–615.

Commons

A frequent mistake in both academic and activist work on land politics is conflating open access resources and commons. It is important to clarify the difference and shed more light on the commons. Commons are 'resources used by groups of heterogeneous users through agreed-upon institutional arrangements' (Agrawal et al. 2023: 532). The defining question is 'how local institutions and communities can viably govern renewable resources such as forests, fisheries, wildlife, pastures, and small-scale irrigation waters' (Agrawal et al. 2023: 532). The term is often used interchangeably with 'open access resources' which are broadly defined as 'resources that are not regulated and are accessible to anyone' (Vaccaro and Beltran 2019: 332). However, 'The commons were not open access... they were a social and conceptual category that included rules and excludability clauses different from individual property or governmental jurisdictions' (Vaccaro and Beltran 2019: 336).

The concept of the commons became one of the most debated themes in social sciences for those with an interest in the politics around resources. This was first triggered by Garret Hardin's article entitled 'The Tragedy of the Commons' (Hardin 1968). The main idea was illustrated by the case of individual herders who use open access grazing land. The assumption was that each individual herder would maximize individual utility by using the grazing land as much as possible, with little or no cost in maintaining it. The result would be the eventual demise of the grazing ground. This assumption was then used extensively to justify and promote the pursuit of individual property rights in land so as to avoid a similar resource collapse.

This seminal article triggered extensive work on the politics of resources and the commons. One of the most important points of disagreement with Hardin is that Hardin was not actually referring to the commons (as defined in the opening paragraph here), but rather to *open access resources*. Note that the definition of the commons is centrally about the institutions governing its access and use. Later, Hardin would concede this point (Hardin 1994). One of the best known scholars who engaged with Hardin, and developed further the notion of the commons, was Elinor Ostrom, whose work earned her a Nobel Prize in Economic Sciences in 2009 (Ostrom 1990). Emphasizing the question of governance of common property resources, Ostrom stressed that users do find ways to set up rules on how to access and use the resource, prevent it from collapsing and generate individual and collective benefits in the process.

> Ostrom [1990] identified a parsimonious set of institutional design principles for the commons: clear boundaries for users and resource systems, congruence between rules and resource systems, rulemaking autonomy, accountable community monitoring, graduated sanctions, mechanisms for adjudication, recognition by higher authorities, and nested institutions.
>
> (Agrawal et al. 2023: 533)

Agrawal et al. (2023) offer a useful description of the different characteristics of resources and goods which helps us to clarify what we mean when we talk about public, common, club, and private goods. They suggest that there are two defining features of resource systems, namely, exclusion (of others from having access to and being able to use a resource) and depletion (how use of resources can lead to depletion or degradation over time). In the study of the commons, these are often referred to as 'excludability' and 'subtractability/rivalrousness'. It is worth including their explanation here:

> Public goods tend toward nonexclusion and nonrivalrousness: It is difficult to exclude people from using them and they do not get depleted with use... Common goods are depleted with use (rivalrous) and it is difficult to exclude users... Private goods... are depleted with use but exclusion is easy. Club goods are available to others even when used by some. Compared with commons, exclusion is easier... Excludability and subtractability are characteristics of goods. But institutions and technologies affect whether exclusion from a good is easy and the degree to which a good is depleted with use. For example, institutions affect whether a resource system is open access or a restricted-use commons. Technological innovation can transform ease of exclusion. The invention of barbed wire, for example, effectively allowed privatization of land in the American West.
>
> (Agrawal et al. 2023: 534)

These types of resources (public, common, club, and private) are distinct from and should not be conflated with, the types of property regimes, namely, public property, common property, open access, and private property. When paired together, elements from resource types and property regime types can produce different outcomes, for example, common resources with open access, common resources with common property, or indeed, private resources with common property, etc. Understanding the distinction between the two elements, but also understanding how they relate to one another, is central in the study of the commons and can avoid confusing usage of terms and the conflation of commons, public goods, and open access resources.

The work of Ostrom and colleagues on the commons expanded and consolidated a global community of scholars who established their own vibrant association and successful journal, and who have been influential on contemporary ideas about the commons. For David Harvey, the preoccupation of Hardin and Ostrom on resources and the commons deflects one's attention from Hardin's original story about private individual herders. Harvey locates the demise of the commons (labourer and soil, invoking Marx) within a different logic: 'Left unregulated, individualized capital accumulation perpetually threatens to destroy the two basic common property resources that undergird all forms of production: the laborer and the land' (Harvey 2011: 106). Following these broad lines, small but animated academic and political conversations have emerged about the commons, but now in the context of struggles within and against capitalism, and in the building of alternatives to capitalism. The concept of commoning is more often deployed by this subset of the academic and activist global community (see 'Commoning' this book).

Recommended reading

Agrawal, A., Erbaugh, J., & Pradhan, N. (2023). The commons. *Annual Review of Environment and Resources, 48*(1), 531–558.
Dell'Angelo, J., Navas, G., Witteman, M., D'Alisa, G., Scheidel, A., & Temper, L. (2021). Commons grabbing and agribusiness: Violence, resistance and social mobilization. *Ecological Economics, 184*, 107004.
Harvey, D. (2011). The future of the commons. *Radical History Review, 2011*(109), 101–107.

Critical Agrarian Studies

The key concepts and how we define them in this book make sense only when understood within the context of the field of Critical Agrarian Studies.[3] Critical Agrarian Studies places utmost importance on how social structures, agrarian institutions and political agency of social classes and groups are constructed, reproduced, and transformed across space and over time. It privileges inquiries into how the exploited and oppressed social classes and groups understand their conditions and try to subvert and change them to achieve greater degrees of fairness and justice, even as 'fairness' and 'justice' are themselves contested concepts. Yet, there are multiple appreciations and interpretations as to what the field is (see, for example, Akram-Lodhi et al. 2021). The view of Edelman and Wolford (2017) has been the first coherent attempt at defining the field. They said:

> Critical Agrarian Studies are simultaneously a tradition of research, thought and political action, an institutionalized academic field, and an informal network (or various networks) that links professional intellectuals, agriculturalists, scientific journals and alternative media, and non-governmental development organizations, as well as activists in agrarian, environmentalist, agroecology, food, feminist, indigenous and human rights movements. These linkages are not easily mapped or bounded, in part because of their complexity and in part because their contours shift over time.
>
> (Edelman and Wolford 2017: 962)

Building on Borras (2023), we define Critical Agrarian Studies as the field that enquires into the causes, conditions, and consequences of societal transformations by focusing its analysis on the interaction between social structures, institutions, and actors that shape the processes of change in, and in relation to, the rural world. It pays attention to questions of agency of the exploited, oppressed, and marginalized groups in the rural–urban and agriculture–industry entanglement, particularly their autonomy and capacity to interpret—and change—their conditions. It is critical in three ways. First, it interrogates theories that champion and justify capitalism and their dominant contemporary variant, that is, neoliberal theories. Second, it is sympathetic to radical social movements and their proposed alternatives, such as food sovereignty and agroecology, but is vigilant in scrutinizing these in theory and practice. Third, it is not only interested in studying the outside world, as it also questions, and works to transform, the very institutions of the global circuits of knowledge: knowledge production, attribution, circulation, and use.

There are three interlocking features of Critical Agrarian Studies today, namely, *political engagement*, *pluralism*, and *internationalism*. These are both as actually existing and as aspirational reference points that together connect the worlds of academic research and practical politics. *Political engagement* means that this field aims to engage in interpreting the world in various ways, in the most rigorous ways, but with the ultimate objective of changing it. Yet the change is not just a random social change; the direction of social change is towards building a more just, fairer and kinder society. Thus, it has a bias in favour of the working class and other exploited and oppressed social groups. This shapes the way scholarship is understood, and the many aspects of scientific work, including the sense of urgency and timing, as well as the medium and format of articulation of research findings. Nowadays, in academia, this is loosely referred to as academic work through societal engagement and with positive societal impact. Work on the politics of land, for example, has to engage with the traditionally exploited, oppressed, and marginalized peoples through their relations to the land and the latter's multiple meanings, and their social movements and allies (see 'Land in Production/Social Reproduction' this book). *Pluralism* is a cornerstone of Critical Agrarian Studies as it is an acknowledgement of the historical roots of the field traced to the two contending ideological forces of the Marxist-Leninist lineage in Agrarian Marxism on the one hand, and on the other hand, the original Russian so-called 'agrarian populists' of the 1860s onwards, through the 1920s of Chayanov, and the 1970s onwards of Teodor Shanin and James C. Scott.

The field has been constructed on the strength of these two traditions, recognizing the validity of each other's criticisms, but trying to navigate and build on synergy while confronting the tensions. This is very much along the arguments made by earlier scholars about the unnecessary polarization between these two traditions, such as the compelling arguments made by Deere and de Janvry (1981) and White (2018), as well as the nuanced analytical reflection by Levien et al. (2018). What is avoided in Critical Agrarian Studies are polarized positions resulting from sectarian tendencies that have characterized the classical agrarian studies debates. *Internationalism* in this field has at least two meanings. On the one hand, it is an acknowledgement that global capitalism spreads unevenly across societies, space, and over time, and thus, the parts can only be understood by looking at the whole, and vice versa. On the other hand, if this is the assumption that requires academic attention, then politically, struggles have to framed accordingly: the national-international dimension, knowing that it is difficult to resolve class struggles around land for instance in one country without addressing its international dimension. This calls for internationalism in working people's struggles. This is partly captured by the slogan from La Via Campesina: *Globalize struggles, Globalize hope!*

The methods in Critical Agrarian Studies centre on the fundamental guiding questions of political economy, focusing on social relations of property, labour, income, and consumption and reproduction, and on how power relations emerge and are contested and transformed. It is firmly grounded in political economy (see 'Political economy', this book) in general, and in class analysis in particular (see 'Class', this book). From there, it branches out widely to intersect with other issues and themes pertaining to social processes, intellectual traditions, fields and disciplines. One of its basic assumptions is that as capitalism penetrates the countryside, processes of commodification of nature (oftentimes, land) and labour lead to social differentiation among the population, making class relations key to scientific inquiry, especially on the politics of land (Lenin 2004 [1899]; Cousins 2024). A very useful and instructive work on methodological challenges in studying dynamics of class, land and labour, and processes of social differentiation is the classic work by Ben White (1989).

Recommended reading

Akram-Lodhi, A. H., Dietz, K., Engels, B., & McKay, B. M. (2021). 1. An introduction to the Handbook of critical agrarian studies. In A. H. Akram-Lodhi, K. Dietz, B. Engels, & B. M. McKay (Eds.), *Handbook of critical agrarian studies* (pp. 1–7). Northampton, MA: Edward Elgar.

Customary land tenure

Across all regions of the world, especially in rural areas, vast lands are accessed and maintained not in and through full-scale individualized freehold private property, but through what is generally referred to as 'customary land tenure'. But what this means can vary from, say, Myanmar to West Africa, from East Africa to the Colombian Amazon. For contradictory and competing reasons, these vast lands are an object of interest for central states, social justice movements and their ally NGOs, land investors, international financial institutions like the World Bank, and multilateral and bilateral aid agencies.

Before we explore different discussions around this concept, it is important to be clear about the basics: how do we define 'customary land tenure'? Chimhowu (2019) offers an extended description that we feel serves our purpose here, and while it is based on the African situation, it also speaks to conditions in other geographic regions of the world. According to Chimhowu

(2019: 898), customary land tenure 'is an omnibus term that at its most basic means collectively owned land usually under the authority of traditional leadership'. The term covers a range of scenarios, from situations in which occupancy of state lands is being permitted to formally registered individual and collective titles within the context of official customary tenure. Traditional chiefs, or large influential local families or clans, or appointed petty government administrators, can act as customary tenure officials deciding who gets to access and use what land, how much, for how long and for what purposes. Some customary conflict resolution mechanisms are also adopted from the norms of the local community (2019: 898). Chimhowu elaborates further:

> Although now much changed, twin key constants of customary lands have been [that] collective or individual use rights are usually earned through 'son of the soil' entitlements. It is however also quite common for both migrants and 'sons of the soil' to pay for such 'birth rights'.... Even though such land may still be collectively owned, individuals can claim rights of use of a specified piece of land through long term occupation and in many jurisdictions this right of use is transferrable to other family members and [in] some part of sub Saharan Africa to those who can pay.... It is not uncommon for a piece of land to have variable use rights across seasons. For example where no long term investment... has been made a piece of land claimed for use by an individual can revert to communal use... during the certain times of the year... Until the neo-liberal reforms, outright sale of such land was not standard practice although informal sales were common.
>
> (Chimhowu 2019: 898)

The global land rush that surged in the early twenty-first century and, later, the rise of climate change mitigation and adaptation politics, separately and together, put the issue of land at the centre of development debates and academic research for reasons related to economic development (economic efficiency of land use and land users) and NBS to the climate crisis (land-based carbon sequestration projects etc.). It is within this broad context that there has been much wider interest as to the fate of customary land tenure internationally (see, e.g., Peters 2013; Springate-Baginski and Kamoon 2021).

In the context of this book, there are a number of key issues related to customary land tenure. First, customary tenure is dynamically evolving. It is tricky to analyse this concept in its ideal–typical form in isolation from empirically grounded observations of the shape of customary tenure in different societies. Chimhowu (2019) offers a useful heuristic tool of differentiating ideal-type customary tenure and actually existing customary tenure, or what we might call customary tenure under dynamic transformation.

Second, because land in customary tenure is held under various levels of informality, that is, anything less than a fully-fledged freehold individualized private property, it is seen by some as an inefficient use of a scarce resource and an obstacle to vibrant market relations. Thus, customary lands have been targeted for privatization and commodification, with land formalization drives being a feature of both the past and the present. Empirical studies have shown that this assumption about lands in customary tenure is not correct always and everywhere—a fact which has been recognized even by some of the original critics from mainstream economics who were advocating for full privatization. Pauline Peters (2004: 270) notes a shift in new institutional economics 'from seeing customary land tenure in Africa as inhibiting agricultural modernization to lauding its adaptive and flexible character that, over time, allows "evolution" towards more efficient forms of landholding'. Ultimately, of course, this tradition in mainstream economics is less attached to the specific institutional form of land reallocation and land market and is more committed to the idea of an economically efficient land use by economically efficient and competitive land users (based on their definitions). It is therefore logical that they would support any tenure system that can offer this.

Third, there are some tendencies among researchers and activists to implicitly suggest, *a priori*, that land-based social relations in customary tenure system are relatively egalitarian. However, careful empirical studies have demonstrated that customary lands are sites of exploitation and oppression. Most of the local communities involved in customary tenure systems are socially differentiated along relations of class, race, ethnicity, gender, and generation. Peters argues that 'the pervasive competition and conflict over land call into serious question the image of relatively open, negotiable and adaptive customary systems of landholding and land use and, instead, reveal processes of exclusion, deepening social divisions and class formation' (Peters 2004: 270).

Fourth, calls for blanket protection through recognition and formalization of customary land tenure—without the elements of redistribution, restitution, and other forms of deep social reforms—can inadvertently result in merely ratifying what exists in these customary lands. In many places, this could mean recognizing and formalizing inequality, exploitation, oppression, and despotism.

The concept of customary land tenure is a catch-all term and thus requires us to be cautious about its usage. It is a mosaic of property relations and property-making processes in which *a priori* generalization is bound to produce puzzling or awkward outliers. It is important to always clarify what we are referring to when we use the term and to be aware of the societal and historical context. It is also important to remember that there can be huge differences between a conceptual assumption of what it is true in one society at a particular conjuncture, on the one hand, and the actual reality in that society and conjuncture, or in another society at a different time, on the other.

Recommended reading

Chimhowu, A. (2019). The 'new' African customary land tenure. Characteristic, features and policy implications of a new paradigm. *Land Use Policy, 81*, 897–903.
Peters, P. E. (2004). Inequality and social conflict over land in Africa. *Journal of Agrarian Change, 4*(3), 269–314.

Notes

1 The below text draws from Borras and Franco (2023).
2 This entry draws on Franco and Borras (2021).
3 This entry draws on Borras (2023).

D

Degradation

'Land degradation is a long-term loss of ecosystem function and service, caused by disturbances from which the system cannot recover unaided' (UNEP's *Global Environmental Outlook* 2007, cited in Andersson et al. 2011: 296). The manifestations of land degradation are found in the form of chemical contamination such as pesticide poisoning, pollution such as that caused by tailings of mines and also include 'salinity, soil erosion, nutrient depletion, overgrazing, deforestation in agricultural landscapes and desertification' (Andersson et al. 2011: 296; Castro-Vargas and Werner 2023). Defined in this manner, the extent of land degradation is clearly significant and has alarmed the international community. There have been various interventions during the past four decades by the United Nations and by key philanthropic foundations, to call public attention to the gravity of the problem and how to address it. Andersson et al. (2011: 296) argued that mainstream studies neglect 'the connections between the different biophysical manifestations of land degradation and its driving forces embedded in a larger societal context'. The authors point out that 'land degradation results from the interaction between the physical environment and society, reflecting social, political and economic processes at different temporal and spatial scales' (2011: 296).

The rise of industrial-chemical monoculture-based agriculture in the context of global capitalism is a key theme in agrarian political economy. The study of this can help deepen our understanding of some manifestations of land degradation such as loss of soil fertility, chemical contamination, pollution, soil erosion and deforestation. The rise of large-scale industrial agricultural production has ruptured the 'metabolic interaction' between humans and the earth. As a consequence, 'it prevents the return to the soil of its constituent elements consumed by man in the form of food and clothing; hence, it hinders the operation of the eternal natural condition for the lasting fertility of the soil' (Marx 1976a: 637–638). This type of production, therefore, 'only develops the techniques and the degree of combination of the social process of production by simultaneously undermining the original sources of all wealth—the soil and the worker' (Marx 1976a: 638; Foster 1999). One outcome of this production system is the

DOI: 10.4324/9781032658438-3

rise of monoculture plantations. Even during its early period, the monoculture plantations struggled to replenish soil fertility eroded by the industrial production method and the long distances between sites of production, consumption and waste (Mintz 1986; Wolford 2021). Profit-making was maintained during that earlier period through continuous forest clearing and the exploitation of natural soil fertility in settler economies, combined with the use of animal bone and guano as fertilizer to sustain profit-making through large-scale production (Friedmann and McMichael 1989; Cushman 2013; McMichael 2013). This was later replaced by mined minerals (phosphate and potash) and synthetic fertilizer (nitrogen) that are tied to the fossil complex. The level of agricultural production achieved in this system cannot be sustained indefinitely, as the supply of minerals and fossils is finite. Global capitalism is the root cause of contemporary environmental problems, including climate change (Malm 2016; Fraser 2021) and soil erosion, such as the dustbowl phenomenon in the 1930s in the United States (Holleman 2018).

In short, it is important to understand land degradation in biophysical terms but equally important to understand its drivers in political economy terms. This is the point being made by some authors coming from political ecology perspective, such as Andersson et al. (2011), and we fully concur.

In addition, we would like to put forward for discussion the idea that political economy and political ecology factors may not only be drivers of land degradation but that their presence may in themselves constitute land degradation even in the absence of biophysical land degradation. We will briefly explain our argument below.

Economic downturn. The agronomic elements and soil fertility of a plot of land may remain intact, but productive activities on that land are no longer economically viable, for various reasons. This happens, for instance, in the context of a commodity boom-bust cycle. For example, lands are planted to coffee, cocoa, rubber or tree nut (e.g., walnut, cashew, macadamia) during a crop boom. During a bust, farmers, especially smallholders, are not able to quickly shift to another crop partly because of the nature of tree crops. While it may be possible to switch nimbly, e.g., from producing annual tomatoes to producing watermelon, this is not an option with tree crops.

Environmental crisis. Land may not be considered fully degraded in biophysical terms, but an environmental event not directly related to the land's biophysical state may render the land essentially degraded and unusable. This can happen in the case of fungus invasion, which has become more frequent in relation to monoculture plantations, such as coffee and industrial banana plantations.

Social deterioration. Land may remain fertile, but a change in the prevailing social conditions may no longer allow smallholder farmers or small-scale pastoralists to use it. For example, an interethnic conflict may render the use of land not close to a village settlement no longer socially safe or desirable, especially when its use generates or aggravates social tension. This is the case between Nu'er and Anoual ethnic pastoralist groups in Gambella, Ethiopia.

Geographic distance. As income from agricultural land progressively declines in many societies, peasants, pastoralists and Indigenous communities increasingly rely on wage work, often in distant locations. Many of them maintain agricultural production in their home village to supplement income from wage work. In some cases, however, agriculture, livestock production or hunting and gathering activities (especially for Indigenous peoples) become significantly less important in terms of livelihood; combined with the problem of geographical distance, this may result in productive agrarian activities being abandoned, even when the agricultural land's biophysical condition is not degraded.

Political strife. Agricultural lands may remain fertile, or grazing land green, but these could become parts of territories that are occupied by hostile armed groups, whether military or paramilitary groups of ethnic forces within the same country, or in the case of civil or cross-border war. To these cases of severe militarization of a territory, we can also add places that have been planted with landmines.

The importance of land both for producing exchange and use value can diminish, or indeed be degraded, even in the absence of biophysical degradation—and the impact on the lives and livelihoods of affected households and communities can be just as profound as biophysical degradation. Therefore, political economic and politico-ecological processes and phenomena should be understood both as drivers of biophysical land degradation as well as constituting as land degradation by themselves.

It is also useful to think of land degradation as *landscape degradation*; here, we can see how the various types of land degradation unfold simultaneously and often co-constitutively with one another. For instance, the boundaries of land degradation can be difficult to demarcate: chemicals used in a plantation may spread beyond the legal boundary of the plantation through water contamination (by river, groundwater), thus causing the spillover of land-degrading chemicals into geographic spaces beyond the plantation (Castro-Vargas and Werner 2023). Similarly, the spreading of pesticides by air may lead to the destruction of natural pollinators such as bees, rendering a nearby orchard degraded in terms of productivity.

Recommended reading

Andersson, E., Brogaard, S., & Olsson, L. (2011). The political ecology of land degradation. *Annual Review of Environment and Resources*, *36*(1), 295–319.

Moreda, T. (2018). Contesting conventional wisdom on the links between land tenure security and land degradation: Evidence from Ethiopia. *Land Use Policy*, *77*, 75–83.

Digitalization

In the 1990s, the Japanese International Cooperation Agency (JICA) collaborated with the Department of Agrarian Reform (DAR) of the Philippines to digitize the DAR's cadastre records by helping set up the Department's Management Information System. By around 2000, most of the information of some 3 million land reform beneficiaries and the land they had been awarded was in digital form, including GPS information on awarded plots of lands. This Philippine land reform digitalization campaign was part of the global trend pushed by bilateral and multilateral development agencies, led by the World Bank. Today, it is part of the broader phenomenon of digital agriculture, and within it, the digitalization of land records.

Digitalization of land records is part of the surge in policy influence of new institutional economics (NIE) that focuses on minimizing transaction costs in all types of markets, which can be accomplished through the provision of complete information. Together with expanded scope of private property rights, especially in land, digitalization is expected to allow for economically efficient production systems and competitive markets to flourish.

The principal mainstream idea is to expand coverage of private property rights in land, preferably on an individual basis, and convert this cadastre record to digital form. Digital forms convey a sense of scientific precision including establishing clear boundaries of a private plot using GPS, avoiding the common problem of boundary conflicts that push up transaction costs in land markets. Moreover, information in the form of digital data can be used extensively and with ease in market transactions, compared to the paper-record equivalents. Thus, digitalization of land is at the very heart of NIE, and its goal is clear: to keep both previously and newly created private property technically secure (no need to worry about paper documents getting wet, or lost, or being destroyed by fire, etc.). This aligns with the digitalization drives in all sectors of the economy: if land is to be used as collateral in a loan, then digital land records will match the digitalization in the banking sector. Massive amounts of data are easy to circulate in bulk in digital form. Finally, land records in digital form can be used in open access databases showing who owns which lands. It is assumed by those who subscribe to the NIE approach that this kind of open access information about land property will lead to transparent and dynamic land markets.

In our view, digitalization of land is neither inherently good nor bad. It does not necessarily and automatically lead to land grabbing by those who can gain access to the information about land. It does not necessarily lead to particular plots of land automatically being integrated into the land market. However, digitalization of land also does not automatically mean securing land rights for working people. Digitalization is a technical process within the deeply political issue of property. In many ways, it is similar to, and overlaps with, the concept of formalization of land (see 'Formalization', this book). In societies where the politics of land is such that only a few control the majority of land, while the majority of ordinary working people have too little land or no land at all, and where digitalization of land is not accompanied by any serious land redistribution programme, digitalizing land could simply mean digitalizing inequality. Digitalization of land cannot be separated from the recently emerged broader politics of digital agriculture (Visser et al. 2021).

Recommended reading

Fraser, A. (2019). Land grab/data grab: Precision agriculture and its new horizons. *The Journal of Peasant Studies, 46*(5), 893–912.
GRAIN (2022). *The digitalisation of land: More data, less land*. Barcelona: GRAIN.
Rotz, S., Duncan, E., Small, M., Botschner, J., Dara, R., Mosby, I., Reed, M., & Fraser, E. D. (2019). The politics of digital agricultural technologies: A preliminary review. *Sociologia Ruralis, 59*(2), 203–229.

Diminution

Dispossession is often interpreted as an either/or issue: one is either in possession of land property or dispossessed of it (see 'Dispossession', this book). It is also often discussed in reference to land for production, and more commonly in the context of property ownership (private, state, commons). Elsewhere in this book, we emphasize the importance of seeing land in the context of production and social reproduction as a connected whole. In this light, we believe that the concept of dispossession can be, or ought to be, accompanied by the notion of 'diminution' in terms of bundle of rights and bundle of powers, or in terms of property and access (see 'Access', this book). The concept of diminution denotes that such a process is a matter of degree, and not an 'all or nothing' issue, and this in turn implies diverse and differentiated types of property rights and land access.

A farming household in a rural area is likely to be inserted in a mosaic of property relations, namely, private, common and state (the three ideal types) and hybrid versions, all at the same time and within the same landscape. For example, a household may have a freehold, privately titled paddy field near the centre of the village, a plot of upland farm inside the village commons, held under a customary tenure in which they only have use rights, and a house plot in the village centre which is state land. In between these plots are spaces that they can use and access, for example, private land which is not being farmed but whose owner allows anyone in the village to graze their goats and cattle on it; a community forest which anyone in the village can access for foraging and gathering non-timber forest products for use and exchange value; a piece of state land in the village centre converted into a public playground for children to play and a community garden for villagers to have their small plots of vegetable production; a riverbank and a spring that are held as commons; criss-crossing spaces and paths in and through private, state and common lands providing right of way to villagers, and so on. In addition, anyone in the village can walk through a private coconut farm, and if they see a coconut that has fallen naturally to the ground, they can gather it. After a farmer has harvested the maize from their private farm plot, anyone can access that space and gather whatever corn cobs are still clinging to the maize stalks or have fallen on the ground. Anyone passing through a private farm plot who sees some fallen branches from a tree can pick up those branches and take them home for firewood. But it is not allowed to pull and gather for private benefit any taro or yam plant from some else's private plot.

In many rural villages in the world, even in the first quarter of the twenty-first century, the hypothetical scenario described above is quite commonplace; even for just one household, the types and layers of land access are plural and diverse. Households are part of a dynamic, fluid property mosaic, in which property does not only mean private property but a combination of ideal-type and hybrid property relations within and across private, common and state property (Macpherson 1978). In other words, we see the co-existence of property relations as both 'bundle of rights' and 'bundle of powers' in the way that Jesse Ribot and Nancy Peluso have convincingly explained (Ribot and Peluso 2003).

In the course of creating this property mosaic, the diminution of rights and powers can occur. The household in our hypothetical case may, for example, lose access to its farm plot under use rights arrangement but retain its freehold private property ownership of the paddy field. It may retain all bundles of rights and powers but lose access to the community forest when the forest is taken over by the state and converted into a commodified REDD+ project under a 'fortress conservation' arrangement. The household may retain all types of land ownership and access, apart from losing a specific right of way when the private owner fences in the area, or access to private land for free grazing for the goats if the land owners decides to withdraw permission, or access to fallen coconuts or postharvest left over maize in the field. In other words, the household in our case may see a decrease over time in the type of rights and powers to access land and resources. In fact, this happens all the time in most places, without the household losing everything all at once, that is, becoming completely dispossessed. Diminution of bundles of rights and powers, of access, is thus a useful concept because it reflects the often partial loss of rights and powers, in contrast to an often all-or-nothing perspective on dispossession. This is especially relevant when we view 'land in production and social reproduction' as a connected whole (see separate entry, this book).

Inherently relational and political, diminution of bundles of rights and powers is neither unilinear nor permanent. It does not imply a progressive accumulation of losses of rights and powers. Rather, such diminution can be multidirectional and subject to flux. Stability in the structure of social relations around rights and powers can only be relative and temporary, as social relations are constantly challenged and contested, and—unlike formal and official cadastre records—never static.

Agrarian justice—the sense of fairness—is, as James C. Scott (1976) explains, quite different from western, especially urban, understandings of justice or injustice. Scott argues that in agrarian societies, peasants tend to view injustice not in terms of how much is taken from them (such as land or crop share) but how much is left to allow a household to subsist. It is this, Scott argues, that determines when and why peasants revolt or support or join revolutionary movements. Following this logic, a carefully disaggregated view of bundles of rights and powers related to land ownership and access, and the concept of diminution of these rights and powers, provides a more nuanced understanding of how much was taken/how much is left, and what kind of political reactions affected villagers take and why. The concept of 'diminution' is therefore useful in understanding agrarian politics better.

Recommended reading

Ribot, J. C., & Peluso, N. L. (2003). A theory of access. *Rural Sociology*, *68*(2), 153–181.

Dispossession

An influential reference point on dispossession is Marx's *Capital* Volume 1 (Part 8) on so-called primitive accumulation (Marx 1976b: 871). The famous lines from Marx on this are as follows:

> In the history of primitive accumulation, all revolutions are epoch-making that act as levers for the capitalist class in the course of its formation; but this is true above all for those moments when great masses of men are suddenly and forcibly torn from their means of subsistence, and hurled onto the labour-market as free, unprotected and rightless proletarians. The expropriation of the agricultural producer, of the peasant, from the soil is the basis of the whole process. The history of this expropriation assumes different aspects in different countries, and runs through in various phases in different orders of succession, and at different historical epochs.
>
> (Marx 1976b: 876)

The history of expropriation of peasants, Marx observed, 'is written in the annals of mankind in letters of blood and fire' (Marx 1976b: 875). Later, David Harvey built on this to argue that such a process does not stop when capitalism develops; rather, it becomes an intrinsic feature of the way capitalism deals with its recurring cycles of crises. From this, Harvey develops his concept of 'accumulation by dispossession' (Harvey 2003). Among those familiar with Marxist political economy, the abbreviations PA for primitive accumulation and ABD for accumulation by dispossession are commonly used (Hall 2013). There are two related factors associated with expropriation or dispossession of peasants in this intellectual tradition: the role played by extra-economic coercion (Levien 2018), and the question of what happens to the dispossessed, that is, whether they are absorbed into the productive sectors of the economy as proletariat or not (see Li 2010). In Marxist terms, those who lose their land property through expropriation but are not turned into the means of production themselves (that is, slaves, etc.) are converted into 'workers' who are free in a double sense: They are free of property and they are free to sell their labour power (Marx 1976b: 874). Another variation of the trajectory in the relationship between accumulation and dispossession, and thus an object of curiosity, is the phenomenon and notion, of accumulation without dispossession (Zhan 2019).

This brief outline of how dispossession is generally understood is deeply embedded in the four fundamental questions of political economy—who owns what? who does what? who gets what? what do they do with the surplus or wealth generated?—or, in other words, questions around property, labour, income and consumption and reproduction. A key point here is that in political economy, dispossession is not interpreted only as a matter of administrative procedure, as for example in concepts such as 'large-scale land acquisition' or LSLA (see separate entry, this book). Rather, the process of dispossession is embedded within the broader imperatives and dynamics of global capitalism. From this fundamental understanding, several related concepts emerge. We discuss some of these here.

First, the general tendency in the literature is to view dispossession from a property-centred perspective which revolves around 'bundle of rights' (Ribot and Peluso 2003). This is important but has some limitations. A more inclusive concept is 'land access' which involves 'bundle of powers' (Ribot and Peluso 2003). The assumption here is that some people may be property-less but may have land access (see 'Access', this book). For example, a household may have no farmland in a village, and so, when a company grabs all farmland in that village, the resulting dispossession will have no relevance to the household members because they were property-less in the first place. However, the members of that household could be dependent on the community forest for part of their livelihood: non-timber forest products for use value and for exchange value. The household has no property claims over any plot in the forest, but it does have existing access. If the same company also grabs the community forest, the household will lose its access. The company's land grab annuls that access and dispossesses this particular household. Thus, dispossession, when framed in the context of Ribot and Peluso's theory of access, is necessarily broader and more encompassing than a formal property-related framing of dispossession.

Second, another general tendency in the literature is to associate land dispossession with the concept of 'land for production'—farmland or grazing area for livestock—usually including settlement sites (e.g. village centres) where people live and have home lots. But for the incessant renewal of life and society, especially in rural communities, land is central to the production/social reproduction connected whole. Here, social reproduction is broadly cast to include not only labour and generational renewal but also socioecological, cultural and political reproduction. In this sense, land is a soil, a productive resource, a landscape and socioagroecological zone or a territory and homeland (see, 'Land in production/social reproduction', this book). Often, land grabs—and therefore dispossession—are considered as such when they occur on lands for production or home lots, and less so when they happen on lands that are largely related to social reproduction (playground for children, village open public spaces, community forest, public spaces like river banks, etc.). This is also a weakness of some methodological concepts such as large-scale land acquisition, as discussed elsewhere in this book.

Third, the implication of the discussion above is that dispossession is often a matter of degree, and tends to be partial and uneven both spatially and temporally. Land access is always plural and diverse in the context of land-in-production/social reproduction. Therefore, unless the land grab is a total grab of an entire landscape or territory or homeland, then it is likely that those affected will be dispossessed of some forms of land access but may retain others. That is why land access is better seen as a 'range of access to a variety of land and nature'. Of course, if land grabbing does involve landscape or territory or homeland grabbing, then dispossession can be absolute (see Hall 2023 for a relevant discussion; see also Tramel 2024.)

There is an extremely useful associated concept, namely, 'regimes of dispossession' developed by Michael Levien. This can be defined as:

> socially and historically specific constellations of state structures, economic logics tied to particular class interests, and ideological justifications that generate a consistent pattern of dispossession. 'Pattern of dispossession' refers to what is being expropriated, from and for whom, to what end, and with what level of success. In brief, in different political economic periods in different places, resources will be expropriated for and from different classes for different purposes that will have uneven ideological acceptance, and which may therefore generate conditions more or less conducive to resistance. By thus periodizing dispossession, light can be shed on what is specific – and perhaps tenuous – about current regimes of dispossession.
>
> (Levien 2013: 383)

The relationship between dispossession and proletarianization is an important subject in agrarian political economy. There are different possibilities in terms of trajectory; we have already mentioned the possibility of dispossession without proletarianization. The surge in the number of the informal sector, or precariat, or relative surplus population (different terms used with overlapping meanings; see Davis 2006; Li 2010) during the era of neoliberalism is partly attributed to the trend of jobless growth amid hyper economic growth in the centres of global capitalism, driven largely by finance capital and labour-saving industries. In the opening decades of the twenty-first century, global land grabbing might have aggravated this situation.

The other major trajectory is the absence of dispossession or only partial dispossession from the land, of rural working people (peasants, pastoralists, etc.) and at the same time their partial absorption into the capitalist profit-making complex as wage workers, usually as part-time precarious workers. A part of the cost of social reproduction of these workers—healthcare, retirement pension, childcare and so on—that should have been included in the calculation of labour cost is then pushed onto the workers and becomes their own responsibility. Seen in a different way, workers are able to survive with low wages because some of the subsistence needs of the household are met through having access to land (farmland, grazing land, community forest, home lot, etc.) back in their rural home village. This represents a vast category somewhere between the ideal-types of the completely dispossessed and fully proletarianized, on the one hand, and the completely dispossessed and not proletarianized on the other. Some refer to this category as semiproletarian, and the process is referred to as semiproletarianization. Relevant readings in agrarian political economy include Karl Kautsky's classic book, *The Agrarian Question*, in which he argued that peasants will be incorporated in capitalism by subordinating and exploiting them but not by dissolving them (Kautsky 1988 [orig. 1899]; see also introduction to the English edition, Alavi and Shanin 1988). On contemporary thinking, see Chambati (2024) and Yeros (2023), as well as Henry Bernstein's notion of classes of labour (Bernstein 2010) and Issa Shivji's rethinking of the idea of 'working people' (Shivji 2017).

Recommended reading

Levien, M. (2013). Regimes of dispossession: From steel towns to special economic zones. *Development and Change, 44*(2), 381–407.

E

Expropriation

> The taking by the state of private property for public purposes, normally without compensation. The right to expropriate is known in some legal systems as the right of eminent domain.
>
> (Oxford Reference Dictionary n.d.)

Expropriation is a keyword, a concept, that was more commonplace during the twentieth century in the politics of land and development than in the twenty-first century. It was associated with state-driven redistributive land reform and rural–urban, agricultural–urban transformations of land in the context of national development campaigns by nation states, especially in developing countries. The keyword 'expropriation' is often used loosely and interchangeably with a number of other terms, namely, 'compulsory acquisition', 'confiscation', and 'eminent domain'. Expropriation is a concept that did not sit well with neoliberalism when the latter surged from the late 1970s onwards primarily because it is anathema to the advocates of voluntary, free market-based principles in running economies and societies. Therefore, it is not surprising that it was one of the concepts that was largely pushed out of the mainstream agenda in development policy and political narratives.

Expropriation is a form of extra-economic coercion used to recast control over land property. As such, where and when it is deployed, it is backed by the coercive apparatus of the state, namely, police, military, paramilitary, courts, and prison. The object of expropriation is private property, taken from its owners in the name of the greater good, or for public purposes; however, the latter is defined. Expropriation has been interpreted and carried out differently within and across societies and over time.

Expropriation usually targets a particular sector of private property owners. In conventional redistributive land policies, such as land reform, it is only rarely the case that a policy officially aims to expropriate all owners of private property. The few examples of this that exist mainly involve socialist revolutions which aim for the complete demise of the agrarian landed class, such as the revolutionary land reform in China in 1949–1952. In the more

DOI: 10.4324/9781032658438-4

common case of capitalist-oriented land reforms, it is only selected sections of the agrarian landed class that suffer expropriation, and on most occasions highly modern and productive capitalist agricultural lands have been exempt.

In redistributive land reform, it is quite common that not all the landed property of affected landowners is expropriated. This can happen in at least two ways. The first method involves the setting by law of a land size ceiling, which means that, for certain types of land, no one can own more than the maximum size decreed. This implies that an expropriated landlord can hold on only to a portion of his estate, the size of which is determined by law. This is also referred to as the landowner's right to retention (see 'Minimum access/ size ceiling', this book). In the second method, the social function of the land is used as the basis for redistribution. This means that, if the land is deemed not to be fulfilling its supposed social function (producing food, generating jobs, etc.), it can be expropriated. This is centrally about the definition and level of productivity of a plot of land. Any plot—whether a portion of an estate or an entire estate—that falls below the benchmark is eligible for expropriation, subject to administrative and legal procedures.

Another contentious issue is the terms of expropriation, which can differ significantly from one society to another. First, there is expropriation without compensation, which is essentially confiscation. This can be done for different reasons. Many socialist revolutions implemented this type. Second is expropriation with compensation. Affected landowners are compensated at different levels, from the full commercial market price for their land, to somewhere below the commercial price level. Some expropriatory land reforms pursue compulsory land acquisition with selective and progressive compensation schemes. This means that for some expropriated land, the landowners will receive zero compensation; the reasons for this might be, for example, that the land was acquired fraudulently, the landowners are engaged in other illegal and illicit business transactions, and so forth. Meanwhile, owners who acquired their land legitimately will be compensated. In general, the bigger the piece of land, the lower the land valuation per hectare; the smaller the estate, the higher the price per hectare. Finally, compensation can be paid in monetary form, usually on an instalment basis, or through government bonds, including bonds for government-owned industries to encourage investments in these strategic industries.

Expropriation is premised on the assumption that the land to be expropriated is for the common good, or with public purpose. This purpose could be direct and explicit, such as to construct a public highway, or establish a new city. It could also be less direct, and could even, at first sight, appear to be fulfilling private purposes, but within the context of advancing the 'social function of property' or a 'social justice' programme.

In the twenty-first century, there have been cases in which beneficiaries of earlier land reforms later had their plots of land expropriated when the state invoked eminent domain to use the lands for the expansion of urban, commercial, and agricultural areas, or the construction of vital infrastructure such as public highways and railways. The widespread conflicts around land expropriation and the terms of expropriation in China, especially in the 1990s and 2000s, are a good example (Andreas et al. 2020). There have also been examples of governments seeking to build special economic zones, for which they need consolidated spaces, as in some places in India (Levien 2018).

In summary, if a government is to carry out redistributive reform, then some elements of a 'minimum access/size ceiling' ought to be put in place; otherwise, the goal to redistribute land is unlikely to happen (see 'Minimum access/size ceiling', this book). But ultimately, if this is the government's agenda and goal, then it can only be accomplished with the use of expropriation as a coercive instrument.

Recommended reading

Oxford Reference Dictionary. (n.d.) https://www.oxfordreference.com/display/10.1093/oi/authority.20110803095805665.

F

Financialization

Financialization is a huge topic. In this book, we are concerned only with the financialization of agriculture and land. It was in 2008 that the global land rush was first reported by the NGO GRAIN (see 'Land rush', this book) (GRAIN 2008). There were many controversial aspects of the land rush. First was the food versus fuel controversy, since many of the reported large-scale land deals were announced by the companies and governments involved to be for the purpose of cultivating feedstock for biofuel. Second, many of the major land acquisitions involved foreign companies from countries that were not conventionally engaged in overseas land acquisitions such as the Gulf states, China, India, Brazil, and South Africa, alongside the traditional European countries and the United States. The third feature was the involvement of companies and sectors that were not traditionally engaged to any significant extent in acquiring and investing in agricultural land, such as car makers and fossil energy companies, as well as financial capital (pension funds, banks, etc.).

Financialization of agriculture and land has been much discussed in conversations around the latest round of global land grabbing. In an important book on this topic, Madeleine Fairbairn suggests that 'we are now witnessing a "financialization of farmland", in which farms are being targeted for finance-sector investment and increasingly valued for their ability to produce financial profits' (Fairbairn 2020: 2). She elaborates:

> Finance-sector investors were drawn to farmland partly for the same reasons as everybody else—crop prices are high, ergo it pays to own a farm—but they also had their own reasons. The year 2008 saw not only a global food crisis but the worst financial meltdown since the Great Depression. As stock prices plummeted and venerable financial institutions teetered on the brink of collapse, farmland, along with other 'real assets', took on a new luster in the eyes of investors. Real assets include such physical investments as real estate, infrastructure, art, and precious metals—in short, assets whose value stems from their substance

DOI: 10.4324/9781032658438-5

> or functions. This is in contrast to 'financial assets'—those intangible assets whose value derives from an underlying contractual claim, such as stocks, bonds, certificates of deposit, and futures contracts. During economic downturns, when the value of financial assets can go up in a puff of smoke, worried investors often buy real assets.
>
> (Fairbairn 2020: 6)

Big international financial capitalists usually establish subsidiaries in target countries; these subsidiaries are then used to capture control of vast tracts of land. This strategy allows companies to avoid regulations related to foreign ownership, but it also allows for the operationalization of agricultural enterprises. This was the strategy adopted by the US-based Teachers Insurance and Annuity Association of America (TIAA) in its Brazil operation (Mendonça and Pitta 2022). TIAA is now the biggest player in the world in this category of investors in agricultural land. It was also the strategy pursued by the traditional agricultural company, Cargill, which established its financial outfit aiming to capture lands in different parts of the world. Cargill also set up a subsidiary in the Philippines for this purpose (Salerno 2014). The financialization of agricultural land has rendered the already complicated politics of land even more complex.

Recommended reading

Clapp, J., & Isakson, S. R. (2018). *Speculative harvests: Financialization, food, and agriculture*. Rugby: Practical Action; Halifax, NS: Fernwood. Open access ebook: https://practicalactionpublishing.com/book/2046/speculative-harvests

Goldman, M. (2020). Dispossession by financialization: The end (s) of rurality in the making of a speculative land market. *The Journal of Peasant Studies, 47*(6), 1251–1277.

Ouma, S. (2016). From financialization to operations of capital: Historicizing and disentangling the finance–farmland-nexus. *Geoforum, 72*, 82–93.

Five Rs

The politics of land is notoriously sectoral. What this means is that land and land use are categorized by the state in different ways (social, economic, political, cultural, etc.), and their governance is subject to different policies and laws as well as different state bureaucracies. For example, farmland usually falls under an agriculture ministry or an agricultural land agency, land restitution is usually handled by a transitional justice agency, Indigenous community land is often the remit of a specific bureaucracy, urban real estate falls under a different office, lands for conservation come under the environment ministry. Social justice movements tend to follow the same sectoral fragmentation: some mobilize around farmland and thus engage politically with an agriculture ministry, others target the environment ministry for conservation issues.

One outcome of this fragmentation is the inability of state bureaucrats and social justice activists to see the totality of land politics or the 'land regime' (this book); another outcome is the frequent deployment by those opposed to social justice reforms of the classic divide-and-rule tactic to weaken popular movements and political mobilizations, or inter-ministry competition and conflict (Suhardiman et al. 2019). For example, the historical tension between Indigenous communities and peasants over competing land claims is often exploited by anti-reform forces to sow political intrigue with the aim of delegitimizing the reform process and weakening the social justice movements (Rojas 2025).

If our agenda is to pursue social justice in and through land politics, it is important to use the broader lens of redistributive and distributive land policies (see 'Redistributive' land policy, this book) as generic frameworks. Using this lens, we can see at least three broad types of land policies today, namely, *redistribution*, *recognition*, and *restitution*. It is quite rare to see serious versions of these three coexisting in one society at the same time. But even if they do and assuming they are implemented successfully, their contribution to positive societal transformation may not be guaranteed. Alongside these three types of policies, we propose to consider two additional concepts, namely, *regeneration* and *representation*. Together, they constitute what we generically call 'five Rs'. Collectively, they aim to respond to the land questions of the twenty-first century. We now sketch each one briefly and explain how and why it is important to see and treat the five Rs in an interlinked manner.

First, where both wealth and the means of production and social reproduction (that is, land) to create wealth, in the rural world, are monopolized by a few, the *redistribution* of wealth and power becomes urgent and fundamental. In the context of agrarian societies, this includes redistribution of access to and control of the key means of production and social reproduction—land, water, seas, and forests—thus contesting the essence of capitalism. Second, where social exclusion, marginalization and discrimination by one dominant social group over other groups constitute an oppressive social complex, the quest for *recognition* has become an important social struggle. In agrarian societies, this can entail recognition of the rights of Indigenous peoples and ethnic minorities over their territory, or women's and young people's independent rights to land (Agarwal 1994; White 2020). Third, where people have lost their land and territory because of broad corporate resource grabs, social struggles for *restitution* represent a key concern. In many rural areas, this means restitution of access to land, territory, water, and forests, especially in light of the global land grabbing of the past decade. This has to be understood as restitution of land in production and social reproduction, which implies the sense of community and the integrity of an agroecological zone within which are individual plots, rather than scattered plots to be restored to individual claimants without consideration of the integrity of the community, landscape or territory. It is important to frame these three together within a common political logic even when there is a plurality of policies and laws for each.

However, even in the unlikely event that that all three policy frameworks coexist and are successfully implemented, land alone is necessary but not sufficient to advance transformational reforms in society. This brings is to the fourth R. Ecological and climate crisis is the hallmark of the contemporary period in world history, largely caused by capitalism and with deep roots in the history of capitalism (Malm 2016; Moore 2017). Social struggles for ecological *regeneration* and environmental justice have become an integral part of broader social justice struggles (Patel and Moore 2017). Moreover, regeneration here is cast in a broader way to include economic regeneration: lands redistributed to rural villagers can easily revert back to the hands of powerful elites without significant support to beneficiaries of redistributive land policies, such as irrigation and infrastructure, and without broader structural transformation of the rural-urban economy more generally. Generational renewal or the reproduction of social categories such as farmers very much depends on the broader dimension of regeneration, as discussed here (see White 2020). Finally, these four goals of deep social reforms can only be accomplished through fierce, relentless and disruptive political struggles within and/or against capitalism that demands *representation* and accountability. Top-down, technocratic style of policymaking and implementation, including well-intentioned ones, often fail because of the lack of meaningful grassroots participation.

The five Rs discussed above are not to be treated as a checklist from which one can cherry-pick. The five Rs are linked in their logic because they are responding to inherently interconnected social processes. This can be seen in a few illustrations. First, contemporary capitalism has seized the opportunity to make profit out of responses to climate change via market-based transactions such as carbon sequestration and trading, biofuels and neoliberal nature conservation. In essence, this is *regeneration by dispossession*, defined here as the capitalist project of operating within and regenerating nature by dispossessing ordinary villagers to facilitate continuous capital accumulation. Many of the conservation initiatives associated with climate change politics are, to varying degrees, associated with this strategy. One way to confront regeneration by dispossession is through *regeneration by restitution*, *regeneration by recognition*, *regeneration by redistribution* or a combination of these, in the context explained above. Many of the problematical land policy frameworks and implementation can be seen critically through this way of cross-sectoral lens, for example, *redistribution without regeneration* marked most of the past and present land reforms programs worldwide.

Redistribution, recognition, restitution, regeneration, and representation can be seen as distinct but interrelated pillars of deep social reforms. The first three Rs can only be pursued if sandwiched by the twin principles of maximum land size (size ceiling) to put a limit to how much land corporations and wealthy individuals can accumulate and a guaranteed minimum land access (size floor) to everyone who would want to work the land. Without such policies, initiatives around land redistribution, recognition, restitution and regeneration are likely to be subordinated to market-oriented policies and social dynamics; demands for land from corporate conglomerates and climate change mitigation and adaptation imperatives will be limitless, while the capacity of other sectors of the economy to absorb villagers who are expelled from their lands is almost non-existent.

While it is important to understand the specific impact of particular land deals, we urgently need to detect and determine, through systematic analysis, the system-wide political impacts of the land rush and climate change politics. These impacts are foundational and far-reaching: they will play a key role in establishing the basic terms and limits of social inclusion/exclusion for years to come, especially in transitional societies. In our view, the political opportunity structure for social justice struggles around land redistribution, recognition, restitution and regeneration has been significantly altered in the contemporary era, making it exponentially more difficult to advance social justice-based reforms in and through the politics of land. But while the challenge for deep social reforms around resource politics has become even harder in the era of the global land rush and climate change politics, the potential for the emergence of broad-based, progressive, multi-class, multi-sectoral, and multi-identity coordinated political struggles has multiplied. These are the kinds of political struggle that are most critical today because they can connect previously scattered sectoral struggles. The kinds of social movement that are needed in the current context are those that are capable of building on and going beyond conventional stand-alone peasant struggles, that is, struggles that combine multiple identities or strategies, e.g., peasants, ethnic nationalities, Indigenous peoples, internally displaced peoples, people uprooted by mines and dams and those that build enduring strategic coalitions between different sectoral movements.

Recommended reading

Rojas, I. (2025). Limits and possibilities of contemporary land struggles by Indigenous Peoples, Black Communities and Campesinxs in the Colombian Amazon. *The Journal of Peasant Studies*, 1–30. https://doi.org/10.1080/03066150.2024.2435557

Suhardiman, D., Keovilignavong, O., & Kenney-Lazar, M. (2019). The territorial politics of land use planning in Laos. *Land Use Policy*, *83*, 346–356.

White, B. (2020). *Agriculture and the generation problem*. Rugby: Practical Action Publishing. Open access ebook: https://practicalactionpublishing.com/book/45/agriculture-and-the-generation-problem

Flex crops and commodities

The global land rush of the twenty-first century had been associated with multiple converging crises: food crisis, fuel and energy crises, environmental and climate crises and financial crisis. The general mainstream thinking is that the solution to these multiple crises lies in the existence of unused or underutilized lands that are available to be appropriated and reallocated for more efficient use to produce commodities that could address the crises—more food, feedstock for fuel, sources of renewable energy such as hydro-power dams, carbon sequestration initiatives and so on. All these require the acquisition and reallocation of land globally. Earlier debates focused on food versus fuel, food versus feed, etc., binaries of the purposes of commodities being produced from the land. However, the realization quickly dawned that this is not an either/or question about commodities, that is, either food or fuel; rather, what is required is the unification of these purposes and uses in a single commodity: vegetable oil for food, fuel and feed. Many of these commodities are essentially 'flex crops and commodities'.

Flex crops are commodities that have multiple uses, such as food, feed, fuel, fibre and other commercial uses of vegetable oil or biomass that can be flexibly interchanged. Flexibility implies the close connections and multiple relationships between different crops and their uses whereby which crop uses can be nimbly switched or substitutions quickly made across a range of crops. For example, sugarcane can produce multiple products with multiple uses: sweeteners of different types, alcohol of different types (for fuel and for rum etc.), and so on. Flexibility means being able to easily switch the final product from sugarcane sweeteners to rum to ethanol for fuel.

Multiple flexible uses of crops are nothing new. What makes current flex crops different is their commercial scale, and the size of the demand for these multiple uses in the global economy. This means that the flexibility of these crops can be a source of profit for big owners of capital. There are a number of variables that can make crops commercially attractive for flexing. Changes in market prices can be an important trigger. For example, if the price of ethanol for fuel is high, sugarcane capitalists may opt to decrease production of sweeteners in favour of increasing production of ethanol. If canola prices are high, palm oil producers may decrease sales of biodiesel and increase sales of food grade palm oil. Changes in policy frameworks can also cause the sudden switch of final products. For example, when the US government increased subsidies and offered special tax reliefs to corn farmers to produce and sell corn as biofuel, the production and sales of corn for food and feed purposes plummeted (causing sharp increases in prices of corn-based commodities and other related commodities), while the production and sales of corn for biodiesel increased sharply.

The commercial viability of any commodity partly depends on the price level at which it can be traded. Owners of capital in the commodity chain of flex crops can produce the commodities at relatively low cost through a number of means. Extractivism, in which nature is converted into factors of production without cost or very cheaply, is one way in which flex crops can be traded for commercially viable profits. For example, clear-cutting forest to open new oil palm plantations has an ecological cost that is not considered and is certainly not paid for by those who stand to profit at various points of the global commodity chain of palm oil. This can also take the form of land grabbing. Extractivism also applies in labour, where the full social cost of labour time, which should include social reproduction costs (healthcare, pension, childcare, etc.) is not paid in full. Extractivism in nature and labour can be achieved partly by capital moving from one geographic site of accumulation to another, usually travelling from core centres of global capitalism to its periphery (most often, the Global South). It can also happen by going across borders to capture cheap labour in the periphery, or allowing and encouraging cross-border migrant wage workers to come to centres of capital accumulation. Migrant wage workers in the global food system, especially in major migration corridors such as Mexico and Central America to the US and Canada, Eastern Europe and North Africa to Europe, Myanmar, and Vietnam to China represent another form of cheap labour extractivism.

Among the best examples of contemporary flex crops are soya (for feed, food, biodiesel), sugarcane (for food, ethanol), oil palm (for food, biodiesel, commercial/industrial uses), corn (for food, feed, ethanol) and trees (for timber, pulp, biomass and carbon sequestration purposes).

Whether these crops are actually used in flexible ways is not always important. At times, anticipatory or speculative narrative is already sufficient to generate the desired effect: to justify investment in and expansion of the production scope of these crops and commodities. Meanwhile, production of these crops is necessarily globally interconnected. For example, corn production in Myanmar surged after 2008. Although it is mainly sold for animal feed, we consider it a flex crop because the upsurge of corn production in Myanmar is connected with the massive transformation of corn use from food and animal feed to ethanol in key production hubs, such as the US Midwest.

Flex crops and commodities have far-reaching implications for how we understand land, land politics, land use, land use change, indirect land use change and crop use change in the world today.

Recommended reading

Agarwal, B. (1994). *A field of one's own: Gender and land rights in South Asia*. Cambridge: Cambridge University Press.

Alonso-Fradejas, A. (2021). 'Leaving no one unscathed' in sustainability transitions: The life purging agro-extractivism of corporate renewables. *Journal of Rural Studies, 81*, 127–138.

Borras Jr, S. M., Franco, J. C., Isakson, S. R., Levidow, L., & Vervest, P. (2016). The rise of flex crops and commodities: Implications for research. *The Journal of Peasant Studies, 43*(1), 93–115.

Food sovereignty

'Food sovereignty' is a scholarly concept that has origins in and also manifests as an important political project by social movements which has influenced international policy processes, albeit unevenly over time and across geographic spaces. It is loosely defined as the right of peoples to construct their own food system in or near their territory, producing and consuming food that is culturally appropriate, healthy and accessible, through ecologically sustainable methods (Patel 2009). It is often contrasted with 'food security', although others see it as more complementary or overlapping (Clapp 2014). It has various components, one of the better-known being agroecology (Rosset and Altieri 2017; Vandermeer and Perfecto 2024)—a closed loop system of production, circulation, exchange, consumption, and waste which, to many, represents an attempt to repair the 'metabolic or ecological rift' (Foster 1999; Moore 2011), and which is increasingly argued to be capable of matching fossil-based industrial agriculture in terms of productivity (Akram-Lodhi 2021). The concept of food sovereignty has not only become popular among food activists both radical and reformist (Holt Giménez and Shattuck 2011), but it has also emerged as a prominent theme in academic research. In addition, several governments have adopted it in their national or subnational policies, albeit with varying interpretations, modes of implementation and initial outcomes.

Food sovereignty, at its core, attempts to dismantle the big capital-dominated, chemical-based, industrial food system while constructing an alternative that represents the opposite: a small-farmer based, nonindustrial, and non-fossil energy reliant food system. In theory and in practice, this can be done through the simultaneous construction and/or transformation of six dimensions, the so-called six pillars of food sovereignty. To fulfil its aims, food sovereignty: (1) focuses on food for people, (2) values food providers, (3) localizes food systems, (4) puts control locally, (5) builds knowledge and skills, and (6) works with nature. In essence, food sovereignty is an attempt to effect a transition in the global food regime.

Within dominant interpretations of the six pillars, the most common tendency in the literature is to locate land in just one of the pillars, usually in 'democratizing resources' as part of pillar #4, 'putting control locally'. Land is commonly interpreted as farmland and, at times, as territory. The normative demand is for agrarian reform and territorial control. This is important, but at the same time, it is limiting both in theoretical and practical political terms. In our view, objectively speaking, each of the six pillars has, or ought to have, a land dimension, and this has to be made more explicit politically and understood conceptually. The implication of this is the need to rethink the dominant view of land in the context of food sovereignty as primarily concerning a productive resource (farmland, grazing land, and so on), and the demand for agrarian reform as principally connected to the sphere of production.

One of our key assumptions, on which we build our argument, is that the politics of land can only be fully understood when seen within the matrix of land/labour and production/social reproduction. We take the spheres of production and social reproduction as inseparable. Moreover, we take social reproduction from a broad perspective, building on labour and intergenerational reproduction at the household level but also going beyond this to encompass socioecological, sociocultural, and sociopolitical reproduction (see 'Production/social reproduction', this book). Labour and intergenerational reproduction is not possible without ecological reproduction, and this is why climate change is widely considered to be an existential threat to humanity, for instance in the form of ubiquitous poisoning of soil in farmland due to sustained use of chemicals and synthetic fertilizers. In multi-ethnic or plurinational societies, land is central to socioecological reproduction; this can manifest in big issues such as Indigenous land reserves or in mundane, everyday matters such as a hill or waterscape considered by a community to be spiritually sacred, or a space for the building of a pagoda, temple or church. Today, societies are built with the idea of a sovereign state and demarcated territory for their sociopolitical reproduction. Thus, land is a part of at least three major categories which are inseparable from one another, namely, resource, ecology and territory (see Ossome 2022; Mezzadri et al. 2024).

We believe that each of the six pillars mentioned above can only be activated through land. Relegating land to just one of the pillars of food sovereignty not only devalues land but also undermines our ability to fully grasp the essence of each pillar. Our view also embeds food sovereignty in actual conditions of working people who need food for their basic survival. This entails localization of food systems (pillar #3) but not necessarily in a singular site. Rather, localization could very well be multi-sited as in the case of migrant workers who split households across multiple sites of production and social reproduction (Pattenden 2018; Shah and Lerche 2020; Mezzadri et al. 2024). For example, in a peasant household in Myanmar, some work their own farmland for half of the year and for the remaining part of the year they work as sugarcane cutters in southern China, bringing infants and toddlers with them but leaving behind school-age children who are taken care of by grandparents in Myanmar (Borras et al. 2022).

This kind of reframing of land politics in the context of food sovereignty gives us a better understanding of how to locate and position various concepts related to land that are urgent and important but are different from or go beyond the conventional 'agrarian reform' demand, even when the latter is tweaked towards a broader formulation. To cite just a handful of examples, these concepts include: 'commons' and 'commoning' (De Angelis 2013; Caffentzis and Federici 2014; Agrawal et al. 2023); land as part of human–animal–nature relations in the context of metabolic and ecological rift and industrial food systems (Weis 2024); land in the context of what Perfecto and Vandermeer call 'nature's mix' in relation to agroecology (Perfecto and Vandermeer 2010; Vandermeer and Perfecto 2024); land as ecology from a world-ecological perspective; land in broad sociocultural and sociopolitical perspectives, to include big concepts in politics like the 'politics of minorization' (Mamdani 2020); land for climate change mitigation and adaptation measure officially sanctioned by UNFCC-related initiatives or as part of centuries-old community practices; and land for kitchen gardens and backyard animal raising, especially in the context of rising peri-urban and urban agriculture (McClintock 2014). Altogether, the discussion so far calls for critical reflection on how we locate and position land broadly within the narrower framework of food sovereignty in theory and in practical politics. It calls for possible reframing of popular advocacy demands such as agrarian reform into broader agroecological zone, or landscape, or territorial reforms, requiring a different understanding of what we conventionally consider a minimum socioeconomically viable farm size, as campaigned for within food sovereignty advocacy. This also requires us to think about the balance between use value and exchange value of land and to talk about the rural and urban continuum, rather than applying an either/or framework. Class relations and how these intersect with non-class social relations such as race/ethnicity, caste, gender, generation, religion and nationality provide a key analytic lens in understanding multilayered social relations around land in the context of food sovereignty. See also 'Land sovereignty', this book.

Recommended reading

Vandermeer, J., & Perfecto, I. (2024). Land and agroecology: Interpenetrating theses. In S. M. Borras Jr. & J. C. Franco (Eds.), *The Oxford handbook of land politics*. Oxford: Oxford University Press, C19S1–C19S13. https://doi.org/10.1093/oxfordhb/9780197618646.013.19.

Foreignization of land

The global land rush of the early twenty first century proved particularly explosive largely because of the involvement of big foreign companies and governments in transnational land investments or what was widely referred to as land grabbing. The South Korean Daewoo land deal in Madagascar involved 1.3 million hectares. It triggered widespread resentment among the people of Madagascar, culminating in sustained protests against the deal (Burnod et al. 2013). The Malibya case involved 100,000 hectares of land in Mali being leased to a Libyan company that had close ties to the Libyan government (Larder 2015). Many of the Gulf states have been directly involved in securing land investment arrangements worldwide, as has the Chinese government. The Japanese and Brazilian governments were at the forefront of negotiations with the Mozambican government to design and implement a massive land-oriented investment corridor called PROSAVANA, that many considered would facilitate land grabs or even constitute a land grab itself (Shankland and Gonçalves 2016; Monjane and Bruna 2020; Wolford 2021). The image of foreign governments and companies taking away lands from ordinary people triggered heated public narratives worldwide about land grabbing; these were at their most intense from around 2008, until media reports about land grabs started to tail off several years later.

It was this image that also helped shape the dominant framework of scholarly research and public debates on land grabs, namely, foreignization of land. The term 'foreignization of land' was more commonly used in Latin America while other regions favoured the term 'land grabbing'. A journal article by Annelies Zoomers in 2010 was one of the earliest papers to put forward a scholarly argument on foreignization of land, suggesting that this concept provided a better analytical framework than the focus on 'in situ processes of land grabbing' (Zoomers 2010: 430). She argued that:

> the global land grab… is causing radical changes in the use and ownership of land. There is a strong growth worldwide of landownership by foreigners, and in many instances the land falls into the hands of long-distance actors (absentees), not those of people from the local community.
>
> (Zoomers 2010: 440)

While clarifying that foreigners buying land is not a new phenomenon, Zoomers explained that the scale and intensity of it around the 2008 global land rush was unprecedented (Zoomers 2010: 441).

Brazil is one of the countries where foreign land investors rushed in. As Sauer and Pereira Leite (2012) have demonstrated, this is not something new for Brazil, but the scale and intensity, as Zoomers argued, were unprecedented. Sauer and Pereira Leite describe the situation:

> The increasing flow of foreign investment in Brazilian land has been met by concerns and protests, from sometimes bitterly opposed social groups. Increasing investments in land assets threaten food security and sovereignty. These investments concentrate agriculture and livestock production even further in a few commodities, and this concentration—both of products and of their ownership—favor monopolies and the increase in the control of the production of food and agro-energy by a few. However, representatives of Brazil's so-called agribusiness as well as opinion articles in the mainstream press, have also voiced alarm over trends of "foreignization". Even as these sectors defend a "pro-market" perspective, there is clearly a voice of concern with respect to the intensity and quantity of land purchases by foreigners. In many ways this presents a far more complex scenario [than] that envisioned.
>
> (Sauer and Pereira Leite 2012: 895)

The concept and narrative of 'foreignization of land' are so powerful not because it emerged in scholarly research but because this was, and remains, the main political sentiment on the ground. It is a nationalist sentiment. National governments reacted to mass resentment against the phenomenon, as described by Sauer and Pereira Leite in the case of Brazil. Different national governments moved quickly to introduce contradictory regulatory frameworks: attracting foreign investors, while trying to regulate foreignization of land, that is, to keep it to a level that was acceptable to the public. As Madeleine Fairbairn observed, 'Placing ownership restrictions on foreigners was a way for the government to respond to this concern and score points with rural social movements, all while burnishing its nationalist credentials' (Fairbairn 2015: 588). Yet, observers were quick to note how relatively easily such restriction were circumvented by foreign investors. As Fairbairn said, 'the simplistic foreign/domestic dichotomy upon which the regulation was based also left substantial leeway for companies to locate financial detours around the law' (Fairbairn 2015: 588).

The Philippine case is interesting in the sense that most of the lands in the southern part of the country that might be of potential interest to particular investors (in this case, Cargill) had already been redistributed to land reform beneficiaries in smaller individual plots. There was also a strict rule against foreigners buying land. Tania Salerno described a particular situation 'where Cargill's private equity arm—Black River Asset Management—is investing in land through equity acquisitions of a Philippine company, Agrinurture, in a manner that allows the company to adapt to national and local dynamics' (Salerno 2014: 1709). This demonstrates that grabbing control of land does not require direct action by a foreign company. The Philippine case is not unique in this sense, as also demonstrated by Fairbairn (2020) in the case of Brazil. Amid financialization of agriculture, food and land (Clapp and Isakson 2018), the connection between grabber and grabbed is not always a simple single line. It could very well be a multilayered, multisited web of interconnections; a transnational land investment web, as described by Borras and coauthors (Borras et al. 2020b), that renders regulation of foreign investments in land less effective.

The dominant framing on foreignization of land has remained influential politically. In Colombia, for example, this political sentiment is only starting to gain momentum at the time of writing, in 2024. In Myanmar, however, the political sentiment against foreign owners of capital—especially from China—has been a long running thread. It echoes public sentiment about land grabbing, especially in the border areas of Northern Shan State and Kachin State where the main method of land control grabbing by owners of capital and diverse land brokers has mainly been through the pin prick style of land accumulation (see 'Land brokers', this book; 'Pin prick', this book). In most cases, land grabbing that involves a foreign entity triggers more political reactions from the general public than land grabs that involve only a domestic entity, even though the source of the capital might not matter much to those directly affected or displaced by the land grab. This requires us to ask some difficult questions that are not meant to downplay the reaction and sentiment of the broader public in affected societies but to try to open up the issues for discussion in political debates. Is a land deal a land grab only when the one who ultimately gets control of the contested land is a foreign entity? What about domestic land grabbers? Which is better, land grabbing by foreigners or land grabbing by domestic entities? For those ordinary villagers directly dispossessed by land grabbing, it might matter less whether a domestic capitalist or foreign investor took their land; the point is, their land was grabbed and they were rendered dispossessed. But for the general public who are not directly affected by a specific land grab, the question of whether the land grabber is a foreigner or not is more important. In this situation, there is a danger of 'ethnicizing' or racializing land grabbing, rather than centring it on the workings of capital.

Recommended reading

Zoomers, A. (2010). Globalisation and the foreignisation of space: Seven processes driving the current global land grab. *The Journal of Peasant Studies*, *37*(2), 429–447.

Forestland

Alongside agricultural land and rangeland that are key elements in studying agrarian transformations and national economic development, forestland is another category that is also substantially about land and land politics. In fact, in the concept of 'forestland', the constant element is 'land' because, as we shall briefly explain below, forestland may not even have 'forest'—the trees, bushes, wildlife and forest people—in it. In Critical Agrarian Studies, we do not take at face value aggregated statistics data, or claim about, forest in forestland; we take it as not unproblematic.

We offer a vignette from the Philippines to illustrate the main ideas we will discuss in this concept. The story is about a big landlord in the region of southern Tagalog in the Philippines. The family of this landlord has controlled thousands of hectares of land planted to coconut trees, a vast monoculture. They recruited tenant–farmers for sharecropping arrangement, where the usual split of coconut harvest is 70–30: 70% to the landlord and 30% to the tenant, but the latter does nearly all the labour. In addition, there are many landless rural labourers who also work for wages for the share–tenants and the landlord especially during harvest and the slight processing needed for harvested coconuts. It is thus, a straightforward agricultural land, and the agrarian social relations between the landlord, share tenants and farmworkers are generally of the semifeudal hybrid. It turned out however that the thousands of hectares of farmland are not officially and properly a private property of the landlord. For in reality, it is a forestland and, as such, a state land. The landlord turned out to have been able to gain control over the vast tracts of land by circumventing a government forest policy in which a maximum of 20 hectares for individuals and 500 hectares for partnerships or cooperatives can be awarded for 25 years renewable for another 25 years for a token annual rental of about 5 US dollars per hectare. The recipients of such a contract need only to maintain the largely forest nature of the land, or at least pursue agroforestry (forestry with some fruit trees). In our case here, however, the landlord converted the forestland into a full-scale agricultural monoculture. This is not a random, unique case. It is quite widespread, and close observers, especially nongovernmental advocacy groups estimated that the forestlands under this kind of actual reality are far too common, quite numerous and if their total in hectarage is aggregated it could be a few millions of hectares.

What is important in the context of this book is that the concept of 'forestland' cannot be taken at face value because the concept has multiple meanings: land with natural forest (the most common assumption of what a forestland is), forestland for conservation (natural forest officially demarcated and assigned the status as for conservation) and forestland for production (usually refer to industrial forestry, natural forest designed for timber extraction), and these can be under institutional arrangement such as national parks, community forests, open access, commons, public land and so on. But the ideal types of forestland can go on: land demarcated and officially categorized as forest but in reality, it is a full-scale agricultural monoculture; officially categorized as forestland, but in reality, it has been long been clear-cut of forest and all timber extracted and the land left to villagers who took over the logged down land and started cultivating it; some forestland might already be semiurban and residential; some forestland are actually cattle ranches or site of an open pit mining. Moreover, a forestland may remain officially categorized as forestland and with actual trees, but in reality, it has been essentially transformed: the natural forest was clear-cut, and replaced by a single species of a fast-growing tree, an industrial forest, on the same block of land. Still, in other places, rubber tree plantation is considered a forest, while in most others, it is not considered a forest.

Forestland is politically contested in their meanings. This is what it means by the term 'political forest' coined by Peter Vandergeest and Nancy Peluso who explained that, "Political forests produce and are products of particular political-ecological relations—congealed and convergent in material, ideological, discursive and institutional relations as well as claims by states or other governing bodies" (Vandergeest and Peluso 2015: 162). They elaborated:

> Contemporary political forests are defined by the scientific, bureaucratic and institutional practices of forestry. They are usually designated, legislated, demarcated, mapped and managed by state forestry institutions, although, more recently, political forests have also been formed, protected and managed by non-state institutions, including conservation organizations, certification bodies or private companies. The practices that define them include establishment of forest land, type and species categories; establishing and empowering one or more professional managerial institutions beholden to the state; and carrying out other forms of forest and forest plantation management. Management practices also involve mapping, patrolling, planning for planting, production and harvesting, managing access within forest boundaries, and compensating persons or other land management institutions displaced by the establishment of political forests.
>
> (Vandergeest and Peluso 2015: 162)

Recommended reading

Vandergeest, P., & Peluso, N. L. (2015). Political forests. In R. Bryant (Ed.), *The international handbook of political ecology* (pp. 162–175). Northampton, MA: Edward Elgar.

Formalization

From the perspective of mainstream new institutional economics, even the poorest of the poor are not asset-less. For example, many poor people in the countryside have access to land. From this perspective, poor people are poor not only because of inefficient land use and land users, but—perhaps even more importantly—because they do not have formal, enforceable claims over assets such as land. Because they do not have formal rights to the land, they cannot use this asset to pursue many aspects of livelihood-making. For example, they cannot go to a formal credit-lending institution like a bank, with lower interest rates, in order to borrow money to fund their production. Instead, they fall victim to informal moneylenders who charge exorbitant interest rates. The lack of formal rights in land property is embedded in a larger maze of informalities: for example, the fact that no formal property taxes are paid means that simple transactions with the government, such as registering one's land claims, involve unnecessarily complex, nontransparent requirements and processes that eventually drive many to resort to informal fixers who charge a fee. This is not conducive to building a vibrant formal economy, without which poor people do not have possibilities for jobs and livelihoods, tax collection is riddled with corruption, and so on. The conclusion is that poor people are poor because much of their capital, or assets such as land, are bogged down by their informal status, and so cannot be used as capital for productive activities. The way to end poverty, and pursue development, is therefore through formalization of institutions related to productive activities. Formalization of land rights, for urban and rural areas, takes centre stage in this framework. The most famous advocate of this idea is the Peruvian economist Hernando De Soto, and his best known book on the subject is *The Mystery of Capital: Why Capitalism Triumphs in the West and Fails Everywhere Else* (De Soto 2000).

Hernando De Soto and his idea travelled far and wide, and he became adviser to many heads of state. Together with Madeleine Albright, Secretary of the State Department of the United States in the Bill Clinton administration, De Soto directed a high-profile commission hosted by the United Nations Development Programme (UNDP), namely, the Commission for the Legal Empowerment of the Poor (CLEP). Its objective was to pursue his ideas and translate them into practical policies and programmes worldwide. The four pillars of CLEP were rule of law, property rights, labour rights and business rights. According to the Executive Director of CLEP, Naresh Singh, poor people 'may be citizens of the country in which they live, but their resources, modest at best, can neither be properly protected nor leveraged' (Singh

2009: 878). Singh highlighted that at the heart of the formalization drive lies the assumption that 'it is not the absence of assets or lack of work that holds them back, but the fact that the assets and work are insecure, unprotected and far less productive than they might be' (Singh 2009: 878) because they are not expressed in formal institutional manifestations. He elaborated:

> Property rights, including tenure security, should not only be protected by law, but also by connecting the property of the poor to wide societal interest (by increasing the range of validation of their tenure security). The possibility is opened for the poor to use property as collateral for obtaining credit, such as a business loan or a mortgage. This encourages compliance by attaching owners to assets, assets to addresses, and addresses to enforcement; that is, making people accountable. As such, property reform can strengthen access to legal identity and to justice. Property records unify dispersed arrangements into a single legally compatible system. This integrates fragmented local markets, enabling businesses to seek out new opportunities outside their immediate vicinity and putting them in the context of the law where they will be better protected by due process and association of cause.
>
> (Singh 2009: 878)

Despite the existence of the CLEP programme hosted in UNDP from 2005 to 2008, the formalization campaign did not accomplish much in terms of practical concrete output. However, it did contribute to legitimizing and widening the reach of the idea that institutional informalities—not capitalist exploitation, oppression and ecological plunder—are the reasons for poor people being poor, and therefore, the argument that the solution is formalization, with land formalization at its heart. CLEP was driving the formalization narrative when the global land rush surged from 2008 onward, further reinforcing that narrative: expulsions and dispossessions happen and integration into capitalist enterprises in large-scale land investments does not materialize because poor people do not have formal rights to their land access. The solution: formalize land rights and access, usually through individual private property rights. Although CLEP was dissolved, land formalization became a catch-all solution to land problems. The subsequent rise in the popularity of digitalization of cadastre records (see 'Digitalization', this book), especially in the era of financialization of land and agriculture (Ouma 2016; Clapp and Isakson 2018), has further reinforced the narrative about informalities as the cause of poverty and land formalization as the strategy to fight against that poverty.

The 2009 critique of this approach by Sjaastad and Cousins is both nuanced and enlightening:

> Current debates around formalising property in the global South highlight a number of specific and undeniable problems, in particular, the

> problem of how to bridge the gulf between customary institutions and those of the state. The objective must be to make state-sanctioned laws and policies a meaningful and stabilising influence in the daily life of the poor rather than a source of opportunism and confusion.... In Africa, supporters of land rights formalisation have a tough time when asked to produce examples of past successes. The general failure of past attempts is at least an indication that formalisation, by itself, is insufficient to reduce poverty. Perhaps the gravest mistake of the 'formalisation at all costs' lobby is their projection of the idea that informal property is the underlying cause of poverty, rendering formalisation an objective in and of itself rather than an instrument that may produce highly variable results depending on its application. What is needed is a wider perspective that recognises the complexities of poverty and its causes, the realities of existing practices and local livelihoods, and the myriad interests and conflicts that often surround land. In short, a recognition of the politics and culture that permeate property rather than its reduction to a technical or legal problem.
>
> (Sjaastad and Cousins 2009: 8)

Advocates of land formalization rarely talk about redistributive land policies. Now and then, the terms 'land reform' and 'agrarian reform' are mentioned in their work, but never explained. There is no discussion about where the lands for which property rights have to be formalized are going to come from or, if the poor are to gain land property, who is going to lose. The absence of any discussion about redistribution or restitution gives us a clearer picture of what the land formalization campaign is and what it is not. Land formalization is ratification of what exists in terms of distribution of land access and control. To formalize what exists: in many societies marked by land-based inequalities, this is essentially *formalizing inequality*.

Recommended reading

Sjaastad, E., & Cousins, B. (2009). Formalisation of land rights in the South: An overview. *Land Use Policy*, *26*(1), 1–9.

Frontier

The land frontier is a relative periphery that is being incorporated into the circuits of capital, either through the extraction of minerals, timber, and other raw materials, or by using land as a productive resource for agricultural and livestock production (farmland, agribusiness, livestock grazing range, etc.). Land frontier-making can either involve small peasants, or large-scale agribusiness. It can be an official, state-driven and organized process with formal institutional rules, or it can be a largely spontaneous movement of individuals going into and clearing the land frontier. Land frontier-making most often transforms land use and land users into capitalist production hubs, although occasionally new land frontiers become part of a peasant economy. In either case, these frontier areas are rarely empty lands; there is a pattern of land use and a set of land users in these spaces, sustained by particular sets of institutional rules, official and unofficial, formal and informal. Thus, land frontier-making implies repurposing of pre-existing land use, and recasting of key actors. The notions of exchange value and use value are then redefined and recalibrated: commodities are produced in these sites, involving inflows of production inputs from outside the frontier (machinery, fertilizers, and pesticides) and outflows of commodities for distant markets. Palm oil, rubber, soya, sugarcane, maize, and fast-growing industrial trees are good contemporary examples.

In an early piece by Jason W. Moore on the concept of 'commodity frontier', he explains:

> The frontier has been such a slippery category because it refers simultaneously to a certain kind of socio-spatial movement and to a certain kind of place—that is, the term 'frontier' refers both to the 'space-of-flows' as well as to the 'space-of-places'. The two dimensions of the frontier mode may be captured in the following formulation. A frontier is a zone beyond which further expansion is possible in a way that is limited primarily by physical geography and the contradictions of capitalism rather than the opposition of powerful world empires. The frontier is a specific kind of space defined by the forward movement of the (capitalist) system. Further expansion is possible so long as there remains uncommodified land, and to a lesser extent labor, 'beyond' the frontier.
>
> (Moore 2000: 412)

Tension and contestation between the structures, institutions and political movers of the old and the prospective new order are standard features in the making of frontiers. The new set of actors wanting to preside over a frontier territory challenge and aspire to dismantle the old order. This often results in wider conflict in frontier areas, as described by Rasmussen and Lund:

> New resource frontiers emerge in different places around the globe. They do not exist as a function of geography per se, but are brought about because new possibilities of resource extraction and use prompt new and competing claims to authority, legitimacy, and access. Frontiers are 'sites where authorities, sovereignties, and hegemonies of the recent past have been or are currently being challenged by new enclosures, territorializations, and property regimes' [Peluso and Lund 2011: 668].... Frontiers are linked to processes of land control and are actively created through social and political struggles. We can push

> this argument further by arguing that a frontier emerges when a new resource is identified, defined, and becomes subject to extraction and commodification.... The frontier moment is a reconfiguration of the conditions of possibility. By unmaking previous orders of property and authority, land and resources are 'freed up' for new forms of appropriation.... The frontier is a contact zone. It is the sites of encounters between knowledge practices, jurisdictions, and visions of modernity, development, and progress. This becomes particularly acute when 'the meaning of place' has become an important object of struggle between central states and subaltern and first peoples.
>
> (Rasmussen and Lund 2018: 391–392)

The making of land frontiers is often associated with spectacle-making: hyperbolic projections about potential windfalls in opening a frontier for whatever commodity production or resource extraction is being anticipated or proposed. Such a conjuncture, where there is much excitement about the prospects of commodity production based on land, where investment money is being generated, and/or a substantial movement of people towards the frontier becomes real—all against the background of spectacularization and hyperbolic projection about huge profits from land investments—can lead to one of two possible outcomes. First, if the government is able to supply enough land through reasonably acceptable institutional mechanisms (rules, procedures, legal documents, etc.), a land boom is likely to emerge. Second, if prospective buyers sense that the government may not be able to supply enough appropriate land, and/or there are no clear institutional rules and procedures for securing the necessary land supply, the result might be a frenzied push—everyone wants to get ahead of the rest, to capture plots of lands. This is a land rush (see 'Land boom' and 'Land rush', this book). Land boom and land rush both have the same result, pushing the land frontier further.

The concept of land frontier was popular in classical agrarian studies and even in the more contemporary Critical Agrarian Studies. Aside from the key terms discussed above, there are two related concepts that are relevant in these fields. The first is Shattuck and Peluso's (2021) idea of everyday forms of territorialization. This comprises the spontaneous actions and movements of individuals who were either expelled from lands near the frontier, or were trying to escape violent conflicts in their communities near the frontier, resulting in them opening new clearings in the land frontier, and informally cultivating the land. The second is a more organized, state-driven process of resettling people in the land frontier as a form of internal colonization. During the twentieth century, this was done by a number of states to avoid redistributing large private estates to poor peasants who were demanding land. Often, then, internal colonization was a land reform-evasion strategy. In many societies, these so-called land frontiers were in fact being used and culturally engaged with by Indigenous communities or other ethnic minorities.

Recommended reading

Moore, J. W. (2000). Sugar and the expansion of the early modern world-economy: Commodity frontiers, ecological transformation, and industrialization. *Review (Fernand Braudel Center)*, *23*(3), 409–433.

Rasmussen, M. B., & Lund, C. (2018). Reconfiguring frontier spaces: The territorialization of resource control. *World Development*, *101*, 388–399.

G

Gender

Henry Bernstein (2010: 115) explained that "class relations are *universal but not exclusive* 'determinations' of social practices in capitalism." He elaborated that class relations "intersect and combine with other social differences and divisions, of which gender is the most widespread and which can also include oppressive and exclusionary relations of race and ethnicity, religion, and caste" (2010: 115). This is relevant in our understanding of the politics of land which is fundamentally grounded in class analysis and interlocking axes of social difference, including gender.

In classical agrarian studies, land is almost always explored in the context of agriculture, pastoralism, and forestry. The treatment is both for productive and social reproductive purposes. However, the dominant and explicit emphasis has been on land for production. Popular themes are land reform, land ownership, land access, land productivity, contribution of agriculture to the national economy, and so on. In this context, gender has become an important issue and has increasingly been mainstreamed, although contested between intellectual traditions. One of the key issues is that methodologies in accounting for national census data, agriculture's contribution to the national economy, cadaster records or landholding distribution data have been dominantly based on the idea of the 'household'. The household used to be assumed, generally speaking, to be headed by men ('the farmer' and 'the pastoralist') who engaged in productive activities that are being captured in formal economy measurements (and the woman is 'the housewife'). One outcome of this is that land policies, especially of the twentieth century had been male-oriented, and to a large extent, excluded women. This was the case for many of the land reforms and land titling programs during that period (Deere 1985; Jacobs 2009). Giving formal land titles only to the men in the households exacerbated the already imbalanced gender relations in the household. During the past three decades, there have been a pushback and has resulted in the subsequent relative popularity of land titles distributed under land reforms that were at least under both the names of the husband and wife, and some land reforms even explicitly recognize the distinct land

DOI: 10.4324/9781032658438-6

claims of women in the household. Two outstanding analytical frameworks and overviews that problematize the household from a gender lens, and thereby give us a better grasp of the fuller dynamics of land politics within broader political economy of agrarian transformations, are Carmen Diana Deere (1995) and Shahra Razavi (2009).

The great revival of theories on social reproduction during the past decade has started to spill-over into the study of land politics, resulting in a much more complex and richer expansion of scholarship on land politics in general, and on the politics of land and gender more specifically. As have emphasized in this book, for us land in production/social reproduction connected whole is key in the incessant renewal of life and society, with social reproduction broadly defined in terms of labour and generational reproduction, as well as socioecological, cultural–political reproduction (see 'Land in production/social reproduction' this book). This builds on earlier land and gender lens but that tended to be social reproduction-focused, in reaction to the production-centred lens (women's access to kitchen garden plots, community forestry for foraging, etc.). The trajectory of the contemporary scholarship, it seems, is towards a lens on land in production/social reproduction connected whole (Mezzadri et al. 2024). This is more complex to study, but it speaks much closer to actual social realities.

Recommended reading

Deere, C. D. (1985). Rural women and state policy: The Latin American agrarian reform experience. *World Development, 13*(9), 1037–1053.

Grabbers

In another entry in this book and building on earlier work, we define contemporary land grabbing as 'the capturing of control of relatively vast tracts of land and other natural resources through a variety of mechanisms and forms that involve large-scale capital that often shifts resource use orientation into extractive character' (Borras et al. 2012: 851; see also Peluso and Lund 2011; Li 2014). This may be 'for international or domestic purposes, as capital's response to the convergence of food, energy and financial crises, climate change mitigation imperatives and demands for resources from newer hubs of global capital' (Borras et al. 2012: 851). There has to be some degree of extra-economic coercion in the process (Hall 2013; Levien 2018); otherwise, it is a simple buy and sell transaction or a hand-out from the state.

One of the concepts that has emerged in the literature on contemporary land grabbing is *land grabbers*. The general tendency in the literature and in political discussions is to treat grabbers as individual actors, either operating alone or in collusion with other actors. Moreover, the focus tends to be on corporate land grabbers, whether they invest in land directly or indirectly, through finance capital, with the state often playing a role as facilitator. One example is the South Korean Daewoo company acting as a land grabber in Madagascar, in a process facilitated by the Madagascan government which would later be the target of citizens' anger (Burnod et al. 2013). The concept of grabbers has contributed to our understanding of who is grabbing land where and what role the state plays in such processes; however, it inadvertently tends to downgrade the role of the state to being a mere facilitator. Several scholars have pointed out the far more active role of the state in masterminding land grabs and taking the lead in all phases of the land grabbing process: espousing the narrative that some lands are inefficiently used or completely unused in order to justify a take-over of these scarce resources to convert them into income-generating enterprises; identifying lands for grabbing; reframing the purposes of those lands; acquiring and finally reallocating the lands. These are the key steps in a land grab process, and it is only the state that has the authority and power to make each of these steps happen. This has been demonstrated by a number of scholars who have analysed the role of the state, e.g., Levien (2018) in the context of India and Lavers (2012) in the context of Ethiopia (see Wolford et al. 2013 for a more general discussion). This highlights the role of the state in a three-way interaction identified by Tsing (2000): capital's globalist dream, the nation-state's notion and assertion of national development and the desire of regional elites to break free from marginalization. The discussion of the role of the state here links back to the concept of 'investment prospecting', paired with 'land prospecting' (see separate entry, this book) in which the state plays the most critical role.

Overemphasizing the role of corporate land grabbers could also inadvertently lead us to undervalue two other aspects of land grabbing: land grabbers engaged in the 'pin prick' style of land accumulation and the layers of 'land brokers' and entrepreneurs (see separate entries, this book). These are important actors driving the process of land grabbing and shaping the trajectory of land grabbing, in particular, and the land rush more generally.

In light of this discussion, we believe that the concept of land grabbers is best understood as an ensemble of actors that, together, makes the grabbing of land control happen, influences, and shapes the trajectory not only of demarcated plots of lands but of the processes of land grab, land boom or land rush more generally. This group is thus always plural in its composition, and always operates as an interconnected whole. One question that is fundamental to understanding the politics of land is the character of interaction among the various parts that make up the whole, that is, the structures that hold the actors, the ensemble, together. This concept of grabbers therefore differs from treating stand-alone actors as land grabbers, be that a company, a government agency, or an individual, and describing the role of each one. By adopting the notion of an ensemble and building on the fundamental assumption that land grabbing is essentially about control grabbing, we have a wider lens through which to examine more complex situations, such as cases of 'Indirect Land Use Change' (ILUC), complex types of land use change and the concept of 'telecoupling' (Lambin et al. 2003; Seto et al. 2012; Meyfroidt et al. 2022) often associated with the rise of 'flex crops and commodities'.

This also gives depth to our understanding of rural and urban dialectics as found in the argument by Serrano (2023). Xu and Borras (2024), for instance, find that while there is rarely any direct land investment from China in soya-producing lands in the Americas, by controlling up to 70% of global soya output through trade, Chinese soya capitalists were able to gain effective control of land equivalent to about 30 million ha. Do we consider this land grabbing? If we include 'contract farming' as a mechanism through which land control grabbing is realized, then why should we not include the concept of ILUC? The point here is that it is important to steer away from a simplistic narrative about individual actors, operating alone or in cahoots with others, and to analyse instead the structure that holds them together—the ensemble as the unit of inquiry. This reveals the more complex concrete social relations that underpin how capital tries to remake its frontiers of expanded reproduction.

Green grabbing

In the literature on contemporary land grabbing and environmental politics, one of the most powerful concepts that has emerged and blazed the trail of an entirely new set of research agendas and political debates is 'green grabbing'. The respected journalist John Vidal coined the term in his 2008 article in *The Guardian* that opened with this paragraph:

> Fancy your own swath of rainforest or snow-capped peak? From Britain to Botswana, the Philippines to Patagonia, there is an explosion of individuals, charities, even billionaire financiers buying up vast areas of land in the name of protecting environments. But is private ownership the way to save them?
>
> (Vidal 2008: n.p.)

Vidal's piece was powerful investigative journalism. McAfee's line of 'selling nature to save it' dovetails with Vidal's point but is situated within academic debates, particularly in political ecology (McAfee 2012). But it was James Fairhead, Melissa Leach and Ian Scoones who best cast the concept in academic terms, constructed the basic frame within which subsequent research and debates have taken place and successfully linked it to basic agrarian political economy (Fairhead et al. 2012). The analytical power of the concept has increased exponentially since then.

For Fairhead et al., green grabbing means 'the appropriation of land and resources for environmental ends' (2012: 238). Central to the concept is 'appropriation'; according to the authors, this 'implies the transfer of ownership, use rights, and control over resources that were once publicly or privately owned—or not even the subject of ownership—from the poor (or everyone including the poor) into the hands of the powerful' (2012: 238). They conclude that 'it is an emotive term because it involves injustice' (2012: 238). It is a subset of the general concept of 'land grabbing', which we define in this book as 'land control grabbing' (see separate entry, this book; see also Peluso and Lund 2011). As such, it is also part of global capital's way of addressing the cyclical crisis of overaccumulation of capital, which requires spatio-temporal fixes to gain control of raw materials for free or at very low cost, and the opening of new markets to achieve the continuous expanded reproduction of capital (Harvey 2003). In this particular case, however, we are also dealing with what James O'Connor (1998) called the second contradiction of capital: that the insatiable appetite of capital to amass ever-increasing profit would run into the ecological limits of such a process. In other words, profit-making, which often treats nature as free or readily available for use as inputs in commodity production, would eventually destroy the ecological basis of the process (O'Connor 1998; Weis 2010). The spatio-temporal fixes (Harvey 2003) fantasize of a win–win formula: addressing the ecological limits and ecological problems brought about by capitalist accumulation, while the 'fixes' themselves can be turned into a new frontier of capitalist expansion and profit-making. Examples include shifting to renewable sources of energy, which has proved to be a lucrative business for the corporate world, and the hijacking of 'organic food and agriculture' by corporate capital (Guthman 2002).

There is a dual narrative that justifies appropriation and reallocation of land in the large-scale land acquisition or land grabbing discourse, namely, that particular types of land users and land uses are economically inefficient and/or ecologically destructive. Either of the two conditions is seen as sufficient to justify appropriation of land, but when the two are fused together, they become even more powerful. Primary targets of this narrative are poor peasant communities, especially those partly or fully engaged in swidden agriculture (Franco and Borras 2019), and small-scale pastoralists, especially those engaged in mobile pastoralism (Scoones 2020). This is the discursive force that underpins much of contemporary land grabbing.

Green grabbing builds on this dual narrative and takes it to a different level. The third element added by green grabbers is the necessity and urgency of redefining and reallocating land use, land access, and land control in the name of a greater good, that is, saving the planet by producing green commodities in the form of renewable energy (hydropower, solar power, wind power, biofuel, and so on), by extracting rare earths and minerals that are necessary materials for green commodities such as electric cars and by constructing carbon sequestration complexes (forest and biodiversity conservation projects and so on) that can be accounted for as carbon commodities for trade, especially in the carbon offset market. Those who are opposed to these projects ostensibly aimed at saving the planet from the climate crisis are seen as obstructionist to be socially ostracized. Green grabbers have thus adopted a tripodal narrative that makes them highly effective in driving contemporary worldwide land grabbing: *economically inefficient* and *ecologically destructive* land uses and land users must be stopped and lands taken away from them, and land use must be shifted to *ecologically regenerating investments* such as hydropower dams, biofuel plantations or forest conservation projects.

Since the publication of Fairhead et al.'s (2012) article, empirical research on specific types of green grabbing has abounded. The literature is largely concentrated on biofuels, or rather 'flex crops and commodities' (see separate entry, this book), as well as on a few key sectors. For solar energy related green grabbing, see Stock and Birkenholtz (2021); for wind power, see Franquesa (2018), Dunlap (2018), and Torres Contreras (2022); for hydropower, see Lamb and Dao (2017); for lithium mining, see Soto Hernandez and Newell (2022); for clear-cutting forest to be replaced by industrial tree plantation, see Scheidel and Work (2018) and Kröger (2014); and for forest conservation, including REDD+, see Ojeda (2012) and Corbera and Schroeder (2011).

Recommended reading

Fairhead, J., Leach, M., & Scoones, I. (2012). Green grabbing: A new appropriation of nature? *Journal of Peasant Studies*, *39*(2), 237–261.

I

Indigenous peoples' land rights

Indigenous peoples' land settlements represent demarcated, formalized and legally constituted areas dedicated to the homeland and territory of particular communities of Indigenous peoples, to ensure their reproduction as a people—not only in terms of labour and generational reproduction but also their social, cultural, and political reproduction. In this context, socioecological reproduction is front and centre of these settlements because in most cases, the reproduction of Indigenous communities depends on the conditions of human–nature relations, that is, on socioecological reproduction (see 'Land in production/social reproduction', this book).

Historically speaking, Indigenous peoples' land settlement initiatives undertaken by central states are relatively recent, but not new: in some countries, they date back several decades. However, since the beginning of the twenty-first century, much greater attention has been paid to Indigenous peoples' issues worldwide, with the United Nations passing the UN Declaration on the Rights of Indigenous Peoples (UNDRIP) in 2006. Accompanying this increased focus on Indigenous peoples' issues has been the rise of the concept of territory, in the particular context of Indigenous communities. In her work on Indigenous community land issues in Bolivia, Penelope Anthias explains that:

> Territory… does not refer exclusively to the state's territory (although this is also relevant), but rather to the modern concept of 'a discrete, sharply bounded and spatially fixed area, premised on spatial exclusivity'… which is 'under the control of a group of people'. This modern concept of territory informs the discursive and cartographic representation of indigenous territories as bounded and contiguous spaces of indigeneity (what I call 'territorially bounded indigeneity'). Territory is thus a utopian idea but one that has political effects, shaping aspirations, imaginaries and practices around indigenous land claims.
>
> (Anthias 2021: 269)

DOI: 10.4324/9781032658438-7

Anthias notes that 'territorially bounded indigeneity imposes a series of restrictions on indigenous claimants, who may be required to demonstrate a special relationship to nature, to a precolonial past and to particular forms of (noncapitalist) development'. She elaborates 'In some contexts, indigenous claimants are required to prove their continuous historical occupation of land and the absence of overlapping claims in order to achieve state recognition' and concludes: 'Indigenous claimants are thus simultaneously called on to affirm their cultural difference and to conform to a modern conception of territory', as distinct from other Indigenous and non-Indigenous groups, conceptually and in practice (Anthias 2021: 270).

The irony in this process is that the demarcation of a specific Indigenous land highlights its inherent relationship to other non-Indigenous spaces and property systems. The Indigenous land settlement is distinct only because of its relationship to all other land property systems and territories outside the demarcated space. This is the essence of Indigenous settlement: the act distinguishes them from non-Indigenous and from other similar Indigenous groups, but at the same time makes an inextricable link to them. The demarcation of an Indigenous land is only legible when viewed in relation to the other spaces from which it is being demarcated, for example, the *fincas* and large agribusiness plantations which surround a formally constituted *resguardo* in Meta department in Colombia. In a related discussion, Mahmood Mamdani talks about a broadly similar process of assigning administrative territories and what he calls the process of 'minorization' or the making of permanent minorities (Mamdani 2020). Our point is that it is impossible to understand the social structures, institutions and politics within Indigenous land settlements if these are artificially delinked from their embeddedness within a broader landscape which comprises a mosaic of property regimes and systems of production and social reproduction which, in the contemporary world, are in a condition of generalized commodification.

The rise of climate change mitigation and adaptation politics since the beginning of the twenty-first century has added to the political currency of Indigenous land rights, with Indigenous peoples being seen as stewards of the land. There is an assumption that Indigenous territories are zones of conserved nature, and thus, a potential and actual site for climate change mitigation and adaptation, especially for carbon sequestration projects. IPCC reports regularly highlight this point, as do many international donor agencies and NGOs. This assumption has much to recommend it: in many Indigenous peoples' territories, nature has indeed been conserved, and many of the remaining tropical forests, savannas, and wetlands in the world today are located in Indigenous peoples' claimed territories. Combined, these amount to a significant area of land/nature and entail social relations that include human–nonhuman relations. This is true whether or not these territories are formally demarcated and legally constituted blocks of land.

However, there are realities that need to be taken into account and potential pitfalls that we should be conscious of. First, many Indigenous communities are yet to get formal recognition from their central states to afford them some kind of official protection against encroachment into their lands. In many places, they are not the only ones mobilizing to take control of these territories; the state, owners of big capital and large conservation organizations also aspire and manoeuvre to claim these same lands. Second, even inside formally demarcated and constituted Indigenous territories, there are instances of formal and informal encroachment by owners of medium and big capital, such as rich farmers, cattle ranchers, owners of extractive activities (artisanal and large corporate mining operations), operators of hydrocarbon extraction, narco lumpen-capitalists and their illicit crop production, and so on. Thus, it is problematic to assume that all legally constituted Indigenous peoples' territories are insulated from capitalist penetration, and therefore, from activities that are harmful to the environment.

Third, owners of big capital do not necessarily oppose demarcated and legally constituted Indigenous peoples' territories. In some cases, they welcome formalized Indigenous land settlements in the context of their capitalist interests: they may be able to reach various kinds of formal agreement with the Indigenous leadership, for example, by manipulating the process of securing 'free, prior and informed consent' (FPIC) by deploying classic divide-and-rule tactics among the Indigenous leaders, in order to pursue capitalist enterprises such as mining operations inside Indigenous lands. In some places, companies even invest in mobilizing the Indigenous communities to gain formal land titles to their land so that they can then forge formal business agreements. In southern Philippines, a major corporation managed to negotiate a long-term (30 year) lease contract covering 30,000 hectares of land in a formally and legally constituted Indigenous peoples' land settlement. In another part of western Mindanao, a big mining company has been able secure an FPIC from a faction of the Indigenous community to pursue a major mining operation (Vidal 2004).

Finally, the rise of nature-based solutions to the climate crisis, REDD+, and other forms of corporate agreement, target the territories of Indigenous communities worldwide. The contradiction is palpable. On the one hand, big business, donor agencies, and NGOs celebrate the stewardship role of Indigenous peoples over land and nature largely because their ways of production and social reproduction are not ecologically destructive. On the other hand, they rush to the Indigenous communities to make agreements to commercialize and commodify these largely uncommodified lands and lives through carbon sequestration and carbon trading business contracts. How this will impact the Indigenous communities in the long run remains to be seen.

There are three more points that are worth reflecting on when we talk about Indigenous peoples' land rights. First, it is possible that many, if not most, of the territories formally demarcated and constituted as Indigenous reserves and settlements are too geographically isolated and of insufficient agronomic and socioeconomic potential to support even the minimum survival needs of the Indigenous communities in those spaces. Arango's study in

Altillanura in Colombia shows that many of those who are officially registered in Indigenous *resguardos* migrate to and from slum areas on the outskirts of small and medium towns in distant locations in search of wage work (Arango 2023). Second, geographic expansion of other social justice- and development-oriented processes, such as land reform, agribusiness, extractive industries and so on, have sometimes been carried out at the expense of Indigenous lands. This has triggered long-standing frictions, including poor-on-poor axes of conflict. In many parts of the world, Indigenous peoples see land reform as an institutional mechanism to grab their land because some of these programmes encroach on their territories. How, then, do we reconcile different strands of social justice struggles when there are serious tensions such as these within the broad justice struggle? Third, formally recognizing Indigenous land rights internationally (as in UNDRIP) and nationally (as in many national laws) is an important institutional milestone in Indigenous struggles for their land and for self-determination. However, what happens in reality is often another thing, and abstract ideas and formal laws do not always sit comfortably together. As Lenzerini wrote in relation to UNDRIP:

> As recently declared by Perry Bellegarde, National Chief of the Assembly of First Nations of Canada, '[t]he adoption of the UN declaration is a milestone in the history of advancing the rights of Indigenous peoples... around the world'. However, the same indigenous leader contextually added that, '[n]ow that we have this commitment, we must work... to realize those rights'. It is the same old story: rights which exist only on paper may be beautiful to read and attractive, but they are not actual rights, because they obviously lack ... effectiveness.
>
> (Lenzerini 2019: 59)

Once again, this brings our discussion back to the key point raised by Ribot and Peluso (2003) on the importance of differentiating bundle of rights from bundle of powers (see, 'Access', this book) and reminds us of Ben Cousin's (1997) question: how do rights become real?

Our main message is this: it is important to see the Indigenous peoples' land issue as a reference point for hopeful alternative ways forward in relation to burning issues in the world today, such as climate change. However, we should also remind ourselves not to romanticize the actual conditions in these Indigenous peoples' territories. A more nuanced and realistic view of how we understand and see Indigenous peoples and their territories is important so that ideas and practices that could contribute to advancing the struggles of Indigenous and working people at large can be further sharpened.

Recommended reading

Anthias, P. (2021). Rethinking territory and property in indigenous land claims. *Geoforum, 119*, 268–278.

L

Land boom

We differentiate the terms 'land rush' and 'land boom'. For us, a land boom is dynamic and energized, but at the same time, it is regularized, normalized, and routinized and accommodated within existing or newly emerged sets of rules and procedures governing its transaction. A schema can be a useful heuristic tool, and in this context, we suggest locating the land boom as a category between '(large-scale) land acquisitions' and 'land rushes'. It shares the element of relatively routinized, regularized, and legitimized processes with the former, but like the latter, it also has some elements of being insurgent, challenging pre-existing limits of procedures, although not to the extent that it generates a convulsive land fever in a generalized atmosphere of madness, driven primarily by spectacle and spectacularization. Land rushes generally cover a much shorter time period than land booms.

This perspective on land boom shares many of the characteristics of a crop boom and its implication for land, as discussed by Derek Hall in the context of Southeast Asia (Hall 2011). Of particular relevance to our discussion here is Hall's notion of 'secure boom' and 'insecure boom', both in relation to land. He suggests that 'in "secure" booms... the basic tenure relations that existed before the boom survive it' (Hall 2011: 839). There are also booms, he notes 'that take place under "insecure" conditions in which pre-existing relations of land control are not respected or accepted by at least some actors' (Hall 2011: 844). He elaborates:

> Insecurity here does not necessarily imply that tenure was insecure before the boom; who controlled what may have been well understood. It does mean that once the boom begins and the value of land rises, these relations are thrown into question.
>
> (Hall 2011: 844)

This is why we find it useful to locate land boom in a schema somewhere between a full-scale land rush and the more everyday phenomena of land grabbing or large-scale land acquisition (LSLA). Within this continuum, a land

DOI: 10.4324/9781032658438-8

boom will usually evolve dynamically along one of two common trajectories: a stable pattern of land accumulation within the overall development of capitalism in a particular region (along lines of the key findings of Tania Li's key findings in her book *Land's End*; see Li 2015) or a boom–bust cycle that many of the crop-associated land booms in history have followed, including those described by Hall.

How then is a land boom related to a land rush in terms of movement along a timeline? A land boom might precede a land rush, implying that it could be part of the build-up towards a full-scale mad scramble for land. But a land rush is often the product of an abrupt insurgent moment. History suggests that a land boom is likely to occur in the aftermath of a short-lived land rush, as for instance in the land boom that followed the land run of 1889 in Oklahoma. During the recent global land rush period, we see this in the land boom that followed the land rush in Gambella, Ethiopia, the land boom that followed the hyperbolic, mad scramble for land in Altillanura in Colombia and the land boom in the aftermath of the land rush in Cambodia—all from 2008 onwards. In Gambella, once the extravagant claims and scramble for land involving both big foreign companies and domestic land investors died down, they were replaced by a steady land market, a land boom, this time involving domestic investors who were land entrepreneurs and aspiring capitalist farmers (Borras et al. 2024). In the case of Colombia, the epicentre of the land rush was the Altillanura region and the northern part of Colombian Amazon (Meta and Guaviare).

Big names in global capitalism, foreign and domestic, were involved in the hype and the overblown projection of the potential of the high plains to become the 'new cerrado' (referring to Brazil's cerrado, a region which has been converted into a massive agribusiness belt). After the spectacle subsided and the frenzy for land dissipated, some large corporations continued to operate, although many of the earlier plans had fizzled out. A land boom then ensued, and the market for buying and selling land in the region has become brisk; these transactions do not always involve big corporations but more of the pin prick type of land grabs (Rojas 2022; Arango 2024). The case of Cambodia is complex but conveys the same message. When the Cambodian government encouraged and facilitated widespread Economic Land Concessions (ELCs), villagers and their allies pushed back through widespread protests. The government responded with repression and concession. The concession part consisted of a nationwide land titling programme in 2012–2013, which benefited small farmers in the villages. The flipside of this was that virtually all lands outside this programme became eligible for ELCs and other business transactions (Dwyer 2015). One effect of widespread ELCs has been the emergence of a land market, and this has been reinforced by the land titling programme. Put simply, many villagers now have individual land titles and can therefore enter the land market and the credit market. After the awards of ELCs and individual land titles, the credit market became extremely active, primarily through microfinance institutions and based on villagers' individual land titles. The credit boom was thus a direct effect of the land boom (Green and Bylander 2021).

Land enclosures, land deals, land grabs, and large-scale acquisitions—and whatever other terms we might use to describe land transaction processes—necessarily go on after a land rush. The waning of land rushes is linked to the routinization of new aspects of the old land regime or the building of emerging commodity and land regimes. It involves the normalization of land acquisitions without the hyperbolic spectacularization and frenzy of a land rush. This transforms meanings discursively and practically so that what were once considered irregular, scandalous and illegitimate land grabs become regular, normal and legitimate land transactions. When this happens, the gathering of data for databases that are dependent on public information about land transactions (media coverage, etc.) is inevitably impacted: key monitoring and measuring mechanisms that depend on media coverage will be weakened because many transactions no longer attract media interest. We believe that what we now see globally is a relatively stable pattern of rules on commodity and land frontier-making which is post-land rushes but part of a continuing land boom. Meanwhile, land grabs go on all the time, with or without land rushes. This requires that the study of land rushes themselves and the study of the relationship between (large-scale) land acquisitions/accumulation, land booms and land rushes should be at least historical in method (Hobsbawm 1972; Bloch 1992 [orig. 1954]; Edelman and León 2013) and preferably comparative.

Land broker

It is important to see both the demarcated plots of land being grabbed and the overall social process of the land rush. In this context, looking into the character and role of land brokers becomes central. 'Land broker' and associated terms are key to our broader understanding of land politics. The concept of 'land broker' is often used interchangeably with other terms in the literature, such as land entrepreneurs, land mafia, middlemen, land consolidators, land scammers and so on. A land broker is someone who makes money not by having land or being a land investor. Instead, a land broker is defined here as someone who makes money by facilitating the conversion of land which has only use value into land with exchange value, or facilitating land sales and rental for a fee or for profit. A land broker therefore has many faces: a formal land sales and rental agent, a 'land flipper' (someone who buys land, makes minimum improvements, invests in promotions and then resells for a windfall), a land document and notary scammer and so on. They can be formal and legal but can also be informal and illegal; they can be government officials or private entities. As such, a land broker is part of the general process of commodification of land under capitalism. The importance of their role in the commodification process is often undervalued, if not overlooked altogether. During a 'land boom', and especially a 'land rush', land brokers of all sorts proliferate and their role becomes more central in the scramble for land as they join a bewildering array of entrepreneurs, prospectors, scammers, and swindlers. The focus of our discussion of this concept is linked to 'land grabbing', 'land boom', and 'land rush' (see separate entries, this book).

In the vast literature on contemporary land grabbing, the discussion about key actors revolves around the 'land grabbed' (the dispossessed) and the 'land grabbers'. For the latter, the focus is on corporate entities, especially transnational companies, and the role played by the state (central and regional). Only very few studies pick up on the critical role played by land brokers and among the excellent ones are Levien (2021), Salerno (2014), and Sud (2014). The concept of land brokers connects our investigations into larger units of inquiry in at least two ways.

First, the concepts of land broker and investment prospector are often connected because land brokers make things happen on the ground. Salerno (2014) gives the example of Cargill, the transnational company, and its search for land for a viable investment in the Philippines. Cargill considered itself too big and too important to deal with landowners offering a few hectares here and there. It needed people who could do the nitty-gritty work for them, acquiring the land it needed, and consolidating this with the lands allocated to Cargill by the government. Without these land brokers, Cargill could not possibly have acquired lands in the Philippines in the locations and at the scale it needed (2014).

Second, a plot by plot, case by case approach to studying land deals, especially when aggregated at a sectoral (e.g., agriculture-related land deals), institutional (buying, renting, voluntary, coercive, and so on), and/or national scale is very useful in our understanding of global land grabs. The bulk of the literature on contemporary land grabbing has followed this approach, and it has radically stretched the boundaries of knowledge on land grabs (Borras and Franco 2024a). In our separate discussion of the concept of the 'land rush' (this book), we emphasize the need to not only study demarcated plots of land at specific conjunctures but also to account for the social processes, that is, the very land rush itself. The aim is to account for a wider range of actors, social processes, outcomes and impacts: direct and indirect, formal and informal, intended and unintended, expected and unexpected. Studies about LSLAs, land grabbing and land booms are important and urgently needed and lead to an improved understanding of specific spatial and sectoral impacts of such land transactions. They require us to study mechanisms of various cases, impacts on social relations (land and livelihoods, etc.) and socioecological entanglements (environmental impact, contribution to climate crisis, etc.). It also requires us to look into the land grabbers, land grabbed, employed or displaced labour, profits, and incomes of land investors and workers or contracted farmers and tax incomes of governments, among others. Studies about land rushes will lead us to the same mechanisms but will include many more besides concerning distant and not immediately related sectors (e.g., the effect on the real estate sector more generally but also on the service sector more widely). They will reveal a bewildering array of actors, from construction workers to mechanics, from food peddlers to all sorts of shop owners and shopworkers (pawnshops, money transfer outfits, telecommunications, etc.), bankers and moneylenders, tailors and dressmakers, carpenters and lawyers, advertisers and bounty hunters, hostels and brothels, transport providers, and all sorts of hawkers, scammers, brokers, and swindlers (Sud 2014; Levien 2021). The impact of LSLAs and land grabs is huge but relatively limited in space and time and restricted to some aspects of social life; the impact of land rushes is far wider spatially, temporally and in terms of social relations, in the context of how land rushes disturb, recast and reshape broader spheres of global social life—food, labour, climate change, citizenship and geopolitics, among others (Borras and Franco 2024b). It is not a matter of which is the more important lens—either LSLAs/land grabs or land rushes. Rather, it is vital that we understand both of these from a relational, multiscale, and multisited perspective and see how they co-constitute one another in particular historical conjunctures (Borras and Franco 2024a).

Recommended reading

Sud, N. (2014). The men in the middle: A missing dimension in global land deals. *Journal of Peasant Studies, 41*(4), 593–612.

Land grabbing

During the past two decades, multiple global crises involving food, fuel, energy, the environment, climate, finance, labour mobility and migration and governance have unfolded from and converged in a common logic: the crisis of large-scale, fossil-based, financialized capitalist industrial development. The symptoms of the crises include the chronic hunger of close to a billion people, although the global supply of food is more than enough to feed everyone; the continuing dependence on fossil-based energy which aggravates climate crises; the increasing role of global finance capital and indebtedness in everyday life (Fairbairn 2014; Gerber 2014; Isakson 2014; Visser et al. 2015); the rise of jobless economic growth, inequality, rural social decay and mass discontent of citizens with their conditions (Davis 2006; Patel and Moore 2017; Scoones et al. 2018; Edelman 2021).

Under contemporary capitalist conditions and assumptions, state and corporate responses to these crises have largely been based on the assumption that, worldwide, there are marginal, underutilized, empty and available lands that can be put to efficient use to produce more food and other commodities, as well as renewable energy. It is assumed that this land can open investment frontiers, generate employment and lead to economic development. This is hailed as a 'quadruple win' for the state, corporations, citizens and the environment (Deininger and Byerlee 2011). This assumption provides the justification for the recent global commodity rushes for food, feed, fuel, renewable energy and carbon through nature conservation. All these require a supply of biomass such as soya, sugarcane, corn, palm oil or trees, thus giving rise to flex crops and commodities (see separate entry, this book) (Borras et al. 2016). Commodity rushes require abrupt and extensive changes in land use and 'land politics' or the process of determining who gets what land and how, how much they get, for what purposes and with what implications. Coinciding with unprecedented urban expansion, and thus, the requirement of capital to convert land uses from agriculture to nonagriculture, commodity rushes have unfolded in ways that included a component of global land grabbing or have unleashed a broader social phenomenon, the global land rush.

The history of capitalism is a chronicle of cycles of land grabbing, beginning with the English land enclosures of the sixteenth century (Marx 1976b, Part 8). We define contemporary land grabbing as 'the capturing of control of relatively vast tracts of land and other natural resources through a variety of mechanisms and forms that involve large-scale capital that often shifts resource use orientation into extractive character' (Borras et al. 2012: 851; see also Peluso and Lund 2011; Li 2014). This may be 'for international or domestic purposes, as capital's response to the convergence of food, energy and financial crises, climate change mitigation imperatives and demands for resources from newer hubs of global capital' (Borras et al. 2012: 851). Based on this 'control grabbing' definition, we see land grabbing as part of the everyday working of global capitalism, validating to a certain extent Harvey's point about the relations between accumulation and dispossession (Harvey 2003). This definition allows us to capture land grabbing that uses different institutional mechanisms, operates at different scales and the outcomes of which are varied. Land grabbing can be corporate-led (or not), large-scale in terms of land and capital involved (or not) and might be undertaken by foreign or domestic investors. In all cases, the general effect is 'control grabbing' of land and resources that, when aggregated at landscape level or society-wide, has the potential to cause shifts in social structures, institutions and politics around land relations. Both large-scale land grabs by corporate capital and the state, and small-scale pin prick style land grabs, are part of the dynamics of a generalized commodification (Hall et al. 2011; Alonso-Fradejas 2012; White et al. 2012; Ouma 2016; Dell'Angelo et al. 2017; Grajales 2021; Neef et al. 2023; Wolford et al. 2024).

One cause of debate in the definition of land grabs is whether extra-economic coercion is a defining element (Hall 2013; Levien 2018). We find the notion of extra-economic coercion too broad and open-ended to play a defining role. For example, a land use zoning ordinance that incentivizes particular crops while disincentivizing others (e.g. eucalyptus versus sugarcane in Guangxi, China), or the objective effect of a mammoth palm oil plantation enveloping most other plots of lands within its radius can, arguably, be considered extra-economic coercion. But we agree that there has to be some degree of extra-economic coercion involved in a land grab; otherwise, it is a plain capitalist buy and sell transaction or a hand-out from the state. Looking at the case of the Russia—Ukraine war, Derek Hall (2023) suggests extending the definition of land grabbing to include territory grabbing by way of military occupation.

The term land grabbing or land enclosure is not new, and neither is the actual phenomenon. But there has been a huge rise in public and academic interest in this issue during the past two decades because of a global spike in land grabs. During this time, different terms have been used to refer to more or less the same phenomenon. The term land grabbing is often used interchangeably with other terms, two of the most common being land rush and LSLAs. However, we believe that a clear definition of key terms is important. Land grabbing, as defined here, is quite different from 'land rush' and 'LSLA (see separate entries, this book). Moreover, even among those who use the term land grabs, there is no consensus on any definition of land grabbing. The definition we offer here, for example, differs from the land size-oriented, proceduralist definition favoured by others, especially those who see a close relationship with the term LSLA.

It is important to clearly define a phenomenon in order to understand what we are investigating and debating. However, it is not straightforward. If we define a concept very narrowly, then we are bound to lose some important elements of the phenomenon. For example, in the 2008–2010 period, the definitions of land grab that were advanced were quite narrow, with the most popular among them having three defining features: big land size, foreign land grabber and a close link to food politics (in the aftermath of the 2008 global food price spike). It did not take long for observers to realize that such a narrow framework failed to recognize many of the actual land grabs that were taking place. But if our definition is too broad, we will lose the distinctive character of this particular cycle of land grabbing in the history and uneven geographies of capitalism; the 'why now and why in particular geographic hotspots?' questions. Peasants and other working people become dispossessed through the everyday differentiating currents of market relations (Lenin 2004 [orig. 1899]). We do not include this kind of dispossession in our definition of land grabbing, although we do include land rush-related 'pin prick land grabs' (see separate entry, this book).

The definition we use here allows us to avoid the pitfalls of too narrow or too broad a definition, and it has three key elements. First, 'control grabbing' is the essential core idea in land grabbing: it is the grabbing of the power to control land in order to derive benefit from this resource (Ribot and Peluso 2003; Peluso and Lund 2011). It is inherently relational and political. This perspective addresses the problem of an approach that is too land-centred. Viewed analytically and empirically through the lens of control grabbing, a land grab does not always require expulsion of peasants from their lands; it does not always result in dispossession (Borras et al. 2012: 849–850). The second defining feature is the scale of the land grab, which means not only the size of land acquisitions but also the scale of capital involved. Taking the scale of capital as the unit of analysis necessarily includes land as central in the operation of capital, while a purely land-centred view (looking at the scale of land acquisitions only) tends to miss or underemphasize the underlying broader logic and operation of capital (Borras et al. 2012: 850). These first two defining features apply more or less universally to land grabs that have happened worldwide, historically. What is distinct about the current era of land grabs—which gives us the third key element in our definition—is that these occur primarily because of, and within the dynamics of, capital accumulation strategies largely in response to the convergence of multiple crises: food, energy/fuel, climate change, financial (where finance capital started to look for new and safer investment opportunities) (McMichael 2012), as well as the emerging need for resources by newer hubs of global capital, especially the BRICS countries and some powerful middle-income countries. In short, the key mechanisms of land grabbing arise from this: food security, energy/fuel security, climate change mitigation strategies and demands for natural resources by new centres of capital. One offshoot of this recent development is the emergence of 'flex crops and commodities' (see separate entry, this book).

Recommended reading

Mollett, S. (2016). The power to plunder: Rethinking land grabbing in Latin America. *Antipode, 48*(2), 412–432.

Oliveira, G. D. L., McKay, B. M., & Liu, J. (2021). Beyond land grabs: New insights on land struggles and global agrarian change. *Globalizations, 18*(3), 321–338.

Peluso, N. L., & Lund, C. (2011). New frontiers of land control: Introduction. *Journal of Peasant Studies, 38*(4), 667–681.

Land in production/social reproduction

The dominant perspective in land policy and politics is to treat land in the context of production, that is, land as a productive resource, for example, as farmland, grazing land, and so on. This is important, but not an adequate reflection of reality. Land is central in production but also in social reproduction, which implies that land means the soil, a productive resource like farmland or grazing area, a landscape, a socioagroecological zone, a territory and homeland. This latter formulation of what land means is based on a perspective of social reproduction that encompasses socio-economic, ecological, cultural and political dimensions, all vital and inseparable in the incessant renewal of life and society.

In classical agrarian studies, the dominant treatment of land is in the context of agriculture which, in turn, tends to be examined to a large extent in the context of the contribution of agriculture to the development of capitalism, or, in some situations, to the construction of socialism (Kautsky 1988 [orig. 1899]; Lenin 2004 [orig. 1899]; see Akram-Lodhi and Kay 2010a, 2010b for an overview; for a recent take, see Shattuck et al. 2023). In this light, land is almost always taken as farmland (or grazing land, commercial forest land and so on) in relation to wider processes of industrialization and urbanization

(Kay 2009). It focuses on the contribution that land makes to the development of productive forces and to producing commodities for exchange or to the process of generalized commodification—in other words, land in the sphere of production. We are not suggesting that this treatment does not include the dimension of social reproduction: it does. But the general focus is on how farmland or pasture or commercial forest land can be used to produce exchange value and use value, and how exchange value can be used in a subsistence-oriented way. This applies especially in the condition of generalized commodification of social reproduction (purchasing food, paying for care services and so on).

We build on this classic formulation, and take a wider view on land, that is, land in production/social reproduction as a connected whole, or land as a means of production and social reproduction, taking a unitary view of the production/social reproduction sphere. Here, we build on Marx:

> Whatever the social form of the production process, it has to be continuous, it must periodically repeat the same phases. A society can no more cease to produce than it can cease to consume. When viewed, therefore, as a connected whole, and in the constant flux of its incessant renewal, every social process of production is at the same time a process of reproduction. The conditions of production are the same time the conditions of reproduction.
>
> (Marx 1976a: 711; see related discussion by O'Laughlin (2022: 1829))

The land in production sphere, that is, land in the production of exchange value, is often expressed in formal census data of land property and land use that are in turn key to understanding the formal capitalist economy. Yet, land is also a means of social reproduction in a broader sense. Access to land, or various aspects of nature at large, such as farmland, food garden, grazing area, forest, water, public space and so on, usually with use value, plays a vital role in the ability of working people (Shivji 2017) or classes of labour (Bernstein 2006) to reproduce their labour, that is, to secure their basic needs for food, clothing, shelter and general care (Pattenden 2018; Shah and Lerche 2020; Ossome 2022). This land access can be substantial in scale and easy to spot, such as access to a community forest not only for the collection of non-timber forest products, but also for the protection the forest offers to the community, for example against storm surges or landslides. But it can also be quite mundane and invisibilized in everyday life, such as a common small pond where a water buffalo can take a regular bath to cool itself, or a public space for a children's playground or for religious infrastructure. A right-of-way may be a non-issue for many, but when you live in a *resguardo* (a Colombian Indigenous People's settlement) which is completely surrounded by private *fincas* and agribusiness plantations, then land access in the form of a right-of-way becomes a key issue that is fiercely contested between those who feel trapped inside the *resguardo* and those who are on the perimeter of it.

Working classes are often able to sell their labour power cheaply, at below survival level, because a large part of their social reproduction cost—healthcare, old age pension, childcare, daily subsistence, etc.—is shouldered privately by the family of the worker (Burawoy 1976). Having access to a small plot of farmland, food garden, or access to forest, water, housing lot, open access rangeland for livestock, community and public health systems allows them to survive and for a few, even to accumulate (Pattenden 2018; Shah and Lerche 2020; Borras et al. 2022). While this is more apparent for rurally rooted workers, even those who are largely or completely urban-based might find themselves in an equivalent position. Access to space closer to sources of livelihoods, whether wage work or street peddling, is key for urban working classes, leading to the mushrooming of slums worldwide. An important section of urban agriculture and animal husbandry is linked to working people's strategy to survive, in which land as urban space proves to be important (McClintock 2014; Siebert 2020; Jacobs 2024).

Since the 1980s, the global economy has produced and invented a wide range of commodities, generated an unprecedented level of wealth, and sustained a food system that is able to produce more than enough to feed everyone in the world. But levels of inequality and impoverishment are equally unprecedented: nearly a billion chronically hungry, about a billion afflicted with food-related health issues such as obesity and diabetes (Swinburn et al. 2019), an industrial food system responsible for one-third of global GHG emissions (Crippa et al. 2021), and at least a billion people who are jobless, underemployed, or with precarious livelihoods, many of whom live in urban and rural ghettos (Davis 2006). This is the conjuncture that has triggered the upsurge of theorizing on social reproduction.

A flash flood, contamination of ground water, depletion of topsoil, severe weather disturbance, destruction of biodiversity, forest clearing—in short, socioecological rupture—can cancel out or make further provisioning difficult if not impossible. If the soil dies due to heavy use of chemicals, productive and social reproductive use of the soil is only possible with artificial support such as even greater use of chemicals: without it, productive pursuits are no longer possible, but with its continued use, the problem only deepens and widens. If pollinator bees are wiped out due to pesticide use that is partly influenced by land use policy, production that relies on pollination will suffer, and social life will be impacted (Kremen et al. 2002). If a community forest which a community relies on for everyday needs such as firewood, foraged food, non-timber forest products for exchange, is clear-cut to make way for a plantation or cattle production, then the capacity of local communities for social reproduction is undermined. Similarly, moments of political–military disturbance, whether due to military occupation of a territory or violent activities of any sort, could destroy infrastructure and impact on the ability of human beings to survive. Some social groups and people (Indigenous peoples, ethnic groups) may not be free to pursue their goal of renewal of life and society if they are prevented from reproducing their *culture*, either through extreme measures like military occupation of their territory or homeland, or through everyday practices of prohibiting aspects of Indigenous and ethnic culture such as religious practices. Moreover, if a big conservation organization deploys a fortress conservation strategy, certain practices of ethnic and Indigenous communities such as swidden agriculture will be prohibited (Heinimann et al. 2017). Indigenous peoples in many societies increasingly complain that they can no longer reproduce many aspects of their culture (younger generations no longer know how to hunt, for example), even—or especially—inside their demarcated ancestral land. In short, renewal of life and society means being able to eat and having a place to sleep but not only that. Thus, production and social reproduction form a connected whole, where social reproduction comprises multiple, inseparable dimensions: socioeconomic, socioecological, sociocultural and sociopolitical.

Recommended reading

Mezzadri, A., Stevano, S., Ossome, L., & Bargawi, H. (2024). The social reproduction of agrarian change: Feminist political economy and rural transformations in the global south. An introduction. *Journal of Agrarian Change, 24*(3), e12595.

Ossome, L. (2021). Land in transition: From social reproduction of labour power to social reproduction of power. *Journal of Contemporary African Studies, 39*(4), 550–564.

Land politics

Land politics is embodied by the set of relational questions around who decides who gets what land, how much, through what institutional mechanisms, why, and for what purposes? This set of questions is closely linked to another set: who decides whether the prevailing status quo in land use, access and control is to be maintained or not, and if not, then what kind of structural and institutional transformations should there be and who are the protagonists that ought to drive and benefit from such transformations?

In this book, we understand land, the object of politics, as social relations around 'soil' (the biophysical being of land), productive resource (farmland, grazing land, land area for extractive industries like gold mining), social reproductive resource (housing plot, public spaces for care and everyday living such as playgrounds, access to community forest, access to a riverbank for doing laundry), territory and homeland, both in a state and nonstate context, especially in relation to Indigenous peoples and ethnic minorities (see 'Land in production/social reproduction', this book). Land politics thus refers to social relations within and between plots of lands, within and between landscapes (see 'Landscape', this book) and territory. Furthermore, we understand land in the context of co-existing and overlapping rights and access or bundle of rights and bundle of powers (Ribot and Peluso 2003) and not as an either/or dichotomy between these two concepts. This implies that the sets of relational questions posed in the opening paragraph are not interpreted only in terms of individual private property but across various property regimes, that is, private, state and common, as well as all possible hybrid formulations within and between these ideal property types.

This perspective on land necessarily requires us to use an intersectional lens when looking at social relations around land and the politics of these. Without class analysis, we cannot even begin to understand the very basics of land politics; but if we only apply a class analysis, we will not get very far with our inquiry. Land-based class relations interlock with other axes of social difference, gender, race, ethnicity, caste, generation, religion and nationality being the most common (see, for example, Agarwal 1994; Deere 1995; Tsikata 2003; Park and White 2017; White 2020).

When we see land in the connected whole of production and social reproduction—as soil, as productive and social reproductive resource, and as homeland—and see land plots, landscape and territory through the intersectional lens of class and interlocking axes of social difference, then land politics become far more complex and multilayered than narrow sectoral conventions suggest. This means that the two sets of questions above are not only understood in terms of individual relations, but as class and social group relations, and in terms of various sectors of society—agricultural, non-agricultural, rural, urban, residential, industrial and so on.

Benedict Kerkvliet (2009) identifies at least three ideal types of land politics. The first is everyday forms of land politics, including how both landowners and ordinary working people engage with one another through overt and covert everyday politics (Moreda 2015), including patron–client relations (Scott 1972) and 'rightful resistance' (Li and O'Brien 2006). The second is advocacy politics, which refers to political mobilizations and actions by peasant movements and their allies such as NGOs to engage in political contests around land. The third, official politics, refers to state policy and politics related to land (Kerkvliet 2009). These forms do not exist independently; rather, they overlap and shape one another. This offers a useful way to understand the various forms and institutional rules through which the politics of land unfold are conducted and transformed.

Recommended reading

Borras, S. M., & Franco, J. C. (Eds.). (2024b). *The Oxford handbook of land politics*. Oxford: Oxford University Press.

Land prospecting

The discussion on the concept of 'land prospecting' in this entry is accompanied by its common counterpart, 'investment prospecting'. Making land investible, as Tania Li (2014) explains, requires that land is transformed from having only use value to producing exchange value. This is the process of commodifying land. Like any commodity being produced and traded in a capitalist society, an element of speculation in generating profit is part of the commodification process. In the context of land, there are two main groups of actors in the commodification process. The first are those who have investment funds, or are brokering on behalf of investment funds, and who are looking for the land needed for a profit-making venture. These are the *land prospectors*. The other group are (state or private) entities who have control or claims over land and are looking for investors who might be interested in that land. These are the *investment prospectors*. When these two actors meet, land transactions can result. This is an everyday dynamic in a capitalist world. But when this process occurs in the context of a land rush, especially a global contagion, then it acquires a different character and unfolds at different scales. The brief discussion here is in the context of the global land rush that unfolded from around 2008.

Speculation and spectacularization are fundamental elements of land commodification, but even more so in the context of 'land grabbing', 'land boom' and 'land rush' (see separate entries, this book). Tsing (2000) identifies three political dynamics underlying these elements, namely, (a) the globalist dream of omnipresent capital, (b) the nationalist aspiration of the nation-state charting its own development trajectory and (c) a subnational regional frontier dream of breaking free from past cycles of marginalization.

In the literature on contemporary global land grabbing, there is a tendency to focus on land prospectors, seen as the land grabbers. And within this set of actors, the emphasis is on corporate prospectors, especially those that are transnational in character. It is suggested that land grabbing generates so much controversy and passion because it involves large foreign or transnational corporations grabbing land across nation-state borders. This emphasis on land grabbers, and on big foreign corporate actors, was especially widespread in the earlier literature of this period, with the 'foreignization of land' a key reference point (Zoomers 2010). National governments in developing countries where large-scale land grabs were happening were projected as 'victims' of profit-hungry foreign corporations. But this phase in the literature did not last long and was replaced by more nuanced studies about domestic land grabbers as well as the role played by central states and regional elites (state and nonstate, including informal chiefs). What emerged was a picture of domestic elites, state and nonstate, who were actively engaged in the process of land grabs (Vermeulen and Cotula 2010; Lavers 2012; Wolford et al. 2013; Levien 2018).

Building on this more nuanced literature, and especially on Tsing's three-way political dynamics among key actors, we offer a complementary way of seeing and understanding these actors by way of the concepts of land prospectors and investment prospectors. The interaction between these two groups and processes is central to understanding the character of land grabbing, land boom or land rush, and some of the elements of these such as 'non-operational and failed land deals' as well as 'pin prick land grabs' (see separate entries, this book). The category of land prospectors—owners and brokers of capital looking for land and commodities—comprises domestic and transnational corporate entities and a bewildering array of land entrepreneurs and brokers. Investment prospectors—those with claims or control over land, or brokers of land, looking for capital investments—very often belong to central and regional state elites (Salerno 2014; Sud 2014; Visser 2017; Clapp and Isakson 2018; Fairbairn 2020; Levien 2021). Once these two groups of actors come together, several factors become relevant. First, both types of prospectors engage in speculation and spectacularization, the process of assuming or projecting that a profit, or even a windfall, can be made from particular investments. The plausibility of such a profit is a necessary element of speculation and spectacle: what is not known, however, is how much land is needed to reach an optimal outcome, and how much investment is necessary to make a viable enterprise. Another important factor is that prospectors are more concerned with making money in the land transaction, than with what is viable in terms of capitalist enterprises; they are focused on the size of the windfall that is promised by the hyperbolic projections. It is also important to understand that prospecting is relational and hypercompetitive. Prospectors are generally plentiful, and they all try to get ahead of the rest. Timing is critical. Finally, existing state and nonstate regulations and procedures such as property rights regimes and investment procedures are not always structured to facilitate the ambitions of the prospectors (Ribot and Peluso 2003; Peluso and Lund 2011; Li 2014). This leads investment and land prospectors to challenge such rules in various ways, including through subversive means. For example, a ban on foreign ownership of land in the Philippines is circumvented by foreign investors setting up subsidiaries or joint venture companies (Salerno 2014).

These unruly processes often result in overblown projections and hyperbolic claims. This is the proximate reason for many failed and non-operational land deals (Borras et al. 2022). The physical, technical and administrative explanations for why some deals fail or become non-operational include difficulty in securing free, prior and informed consent from local communities, unfavourable socioecological and agronomic conditions, difficulty in aggregating small plots into a larger whole and so on. However, factors related to politics and the political economy of land and investment prospecting in the context of land boom and land rush are equally important. Building on Tsing (2000), we argue that the concept of prospecting and the roles of land prospectors and investment prospectors have to be understood in a relational manner.

Land reform

In the classic, conventional sense, land reform means the net transfer of land-based wealth[1] and power from the landed to landless and land-poor classes. Griffin et al. (2002: 279–280) define land reform as redistributing 'land ownership from large private landowners to small peasant farmers and landless agricultural workers', emphasizing that it is 'concerned with a redistribution of wealth'. From a social justice perspective, it is generally understood to correct some historical injustice; from an economic perspective, it is judged to rid an economy of a major barrier to economic development, namely, feudal or semifeudal landlordism.

According to Tuma (1965: 251), land redistribution aims to create 'purposive change' that can result in the improvement of the situation of the landless and land-poor peasants and rural workers. Such purposive change or 'reform' is inherently relational: it must result in a net increase in poor peasants' and rural workers' power to control land resources with a corresponding decrease in the share of power of those who used to have control over the same land resources and production processes. Land redistribution is essentially power redistribution. This can occur through the transfer of the entire bundle of property rights, including the right to alienate land, but it can also be realized without involving full, formal ownership, for example, through leasehold reform and stewardship (see Herring 1983: 13; Putzel 1992: 3; Byres 2004: 27–32). Thus, what is meant here by land reform is not a politically neutral change in production and distribution relationships in a given agrarian structure. The latter ('change') can happen in multiple directions both within and between social classes, as it may include elite-to-elite or even poor-to-elite transfer of effective control over land resources. The former ('reform') limits the direction of change to that which transfers power between social classes, specifically, from landed to landless and land-poor classes or from rich to poor. The redistributed lands can be held either collectively or individually, while the organization of production can take the form of family farm or corporate-type plantation, and historically has been pursued from either a capitalist or a socialist perspective.

Redistributive reform is a matter of degree. The redistributive nature of a land reform transaction in a given landholding, and the change that it causes in the relative shares between the landlord on the one side, and the peasants on the other side is seldom either 100% redistributive or completely non-redistributive; it is usually somewhere in between. It is rarely a purely 'zero-sum' process. The extent to which redistributive land reform is implemented in any society is also a matter of degree, with redistributive outcomes seldom being all or nothing. Traditionally, two interrelated elements have defined the redistributive character of a land reform policy, namely, the compensation to landlords and the payment made by peasants. On the one hand, compensation to the landlord can be between zero and somewhere below the 'market price' of the land; the difference between the market price and the actual compensation partly defines the degree of redistribution. On the other hand, the payment made by peasants and rural workers for the land can be between zero and somewhere below the acquisition cost, with the difference between the two also partly defining the degree of redistribution. Compensating landlords at full market price and asking peasants to pay for the full cost of land acquisition is nothing but a land sale transaction brokered by the state acting as real estate agent. We do not consider it here as a proper land reform in the social justice tradition (see 'Market-assisted land reform', this book).

Taking redistributive land reform as inherently a matter of degree provides us with an analytical tool to understand and compare land reforms between and within countries. Using this perspective, analysis can move beyond the crude 'success' or 'failure' dichotomy, which is also overly quantitatively oriented, and bring qualitative aspects into the analysis to allow for more nuanced comparisons, especially on the social and political–economic aspects of land reform. For example, a land reform that confiscates lands without compensation to landlords and distributes such lands to peasants and workers for free constitutes redistributive reform. Similarly, a land reform that expropriates lands with compensation to landlords at below market price and distributes such lands to peasants and workers at reduced or subsidized cost is also redistributive. However, the degree of redistributive reform is higher in the former than in the latter. Such is the case comparing land reforms, by both nature and extent, in China and Taiwan immediately after World War II (see Griffin et al. 2002). A more nuanced perspective also allows for a better subnational analysis of land reform: the surge of redistribution of irrigated private lands in northern Mexico in the 1960s is arguably more redistributive in nature than the redistributions of many marginal public lands in the central part of the country in the 1920s (Sanderson 1984). The land reform achieved during the Allende administration in Chile was, by nature and extent, more redistributive than that during the Frei administration (Kay and Silva 1992). The difficulty of devising a tool to measure the degree of redistribution should not be underestimated, which makes statistical generalizations on land reform outcomes challenging, but this does not weaken our claim that land redistribution is a matter of degree.

The two minimum requirements for redistributive land reform, namely, compensation to landlords at below market price and payment by peasants and workers at below actual acquisition cost, must, in turn, be linked to the principle that land is not a purely economic factor of production. Rather, land has a multidimensional function and character, with political, economic, social and cultural dimensions. The 'value' of land cannot be reduced to strictly monetary terms, and so the 'market price' of a parcel of land is actually a contested notion involving political–economic and socio-cultural factors which themselves depend on who is attaching the value to the land. The notion of land having a multidimensional character provides the basis for including in the analysis issues which are imbued with value judgements, such as 'social justice', 'social function of the land', 'purposive change' and 'empowerment', which cannot be understood in purely monetary terms. It also highlights the role of the state in the interventions needed to achieve the multiple goals of land reform policy. Thus, landlord compensation-related mechanisms are also largely determined and should be determined, by non-economic factors such as sociohistorical circumstances and the politics of pre-existing land monopoly and reform. The same consideration applies in determining the level of payment by peasants and workers. Hence, the multidimensional character of land renders the monetary-based valuation method an important but incomplete way to assess the actual and full value of the land.

The scope of a land reform is also a matter of degree and differs from one reform programme to another, both within a country and between countries. For example, the land reform programme in Brazil that started in 1988 is based on the social function of the land; through a constitutional provision; the goal of the programme is to expropriate estates that fall below a minimum level of production or productivity (that is, not performing their social function). The principal target landholdings are thus the idle lands used for speculation or lands used in land-extensive cattle ranching. In contrast, the 1988 Philippine land reform law, applying constitutional provisions under social justice, declared that all lands, regardless of size, production and productivity would be redistributed to peasants, with landlords having the right to retain only 5 hectares of land. Of course, how the two land reform laws were actually implemented varied over time and in different subnational spaces, but the fact remains that these are two progressive land reforms laws, which have significant differences in terms of scope.

Finally, while conceptual clarification about redistributive land reform is crucial in understanding the nature and implications of a land reform policy, confusion can arise among scholars and policymakers regarding the empirical data they are working with. Specifically, changes in who supposedly controls land might be registered in formal, official records, but not actually occur in reality. Owners of land are notorious for keeping tight control of lands but often keep official records that show multiple owners of the land, many of which are 'dummy' owners. The conceptual clarification that land reform is about redistribution of wealth and power necessarily implies and requires that analysis of land reform must investigate actually existing conditions rather than relying uncritically on what the official statistical data claim or convey. This is because social relations that are essentially power relations—the very subject of reform—cannot be fully and properly captured by static official statistics alone.

Recommended reading

Griffin, K., Khan, A. R., & Ickowitz, A. (2002). Poverty and the distribution of land. *Journal of Agrarian Change, 2*(3), 279–330.

Wolford, W. (2010). *This land is ours now: Social mobilization and the meanings of land in Brazil*. Durham, NC: Duke University Press.

Land regime

In a situation of generalized, although uneven, commodification of production and social reproduction, land is hurled to the front and centre of commodity production and profit-making. Land is a means of production and social reproduction but also a sink to absorb production waste such as carbon. In this context, it is important to think about land—whether as a soil, a productive resource such as farmland or grazing land, a landscape or socioagroecological zone, a territory and homeland—in relation to global capitalism. This is not to suggest that capitalism determines land politics in an absolute manner; rather, it is to suggest that capitalism is the single most important social force that shapes the meanings of land in particular places and conjunctures. If global capitalism is the broader system within which land is incorporated at the contemporary conjuncture, then it is important to think about land at the same scale as global capital, that is, systemic and global with particular manifestations and characteristics and not only as a plot of farmland or a piece of real estate. Here, the concept of 'land regime' is relevant.

By land regime, we mean the relatively stable pattern of interconnections between land-based social structures, formal and informal institutions that enable, provide and maintain the scaffolding for such structures, and the correlation of state and social forces and political power that legitimize this structural and institutional order. We build this concept from existing studies that take pluralist approaches to land property systems, including Agrawal et al. (2023), Boone (2013), Levien (2018), Lund (2006) and Ribot and Peluso (2003).

It is important to study dynamics of land use (e.g., monoculture oil palm, residential-commercial real estate or conservation), production systems (e.g., smallholder farming, industrial plantation, special industrial zone) and socio-ecological relations (how land use and production systems transform nature). But it is equally important to study the class formation and constellation of social groups that operate as cause and effect and reproduce particular patterns of land use, production systems and socioecological relations. The assemblage of agribusiness, mining and cattle capitalists, in alliance with narco-capital, finance capital, commercial–residential real estate elites and big conservation organizations, may seize and apportion among themselves rural and urban spaces, land and water, in ways that may or may not be fully sanctioned by the state. This may result in a reduction of land and water access for sections of society, pre-empt future redistribution and restitution, dispossess rural working people and lead to the emergence of a particular labour and livelihood regime. If we are to understand the spatial and temporal dimensions of land use and production systems, it is crucial that we examine the class formation and interlocking axes of social difference that make up social structures. A focus on the overall structure does not mean losing sight of the parts; rather, it reminds us to view the parts in relation to the whole, that is, seeing farmland, fish farms or pig farms, residential subdivisions, grazing land and mining sites not as unrelated, random spaces and land uses, but rather as particular manifestations of a whole system.

Class relations, and thus social structures, evolve in cultural context and are inherently historical (Thompson 1991 [orig. 1963]). Their character and trajectory in particular landscapes and at certain conjunctures are shaped by state and nonstate, formal and informal institutions (procedures, rules, norms), however contradictory these might be (Steinmo et al. 1992: Chapter 2). When a state declares spaces to be 'wasteland', i.e., unused and uninhabited, this is an institutional artifice that entices internal colonizers to seize lands, often from indigenous and other occupants or land users, or enables the state to either sustain a prevailing land-based social structure or reallocate land use and thereby trigger change. Armed groups occupying territories can prohibit or allow mining companies to extract resources or agribusiness to produce commodities, which may complement or run counter to formal state plans and regulations.

Furthermore, the correlation of state and social forces and their political power in favour of, or in opposition to, a particular institutional set-up which either maintains or subverts land-based social structures can make or unmake a land regime. This can be done by promoting specific kinds of economic relations, or deploying extra-economic coercion, or combining the two. The widespread dismantling of farmland collectives in China, Vietnam and eastern Europe in the 1980s and 1990s is one example (Kerkvliet 2005; Ye 2015). A central state may pass a law on peatland conservation, for example, but the interpretation and implementation of laws depend on the balance of state and social forces for or against such a law.

Under generalized commodification of everyday life, using a landscape lens to observe and study the role played by land in the expanded reproduction of capital reveals a fuller interaction between structures, institutions and power. Land is central to global capitalism's ability to continuously reinvent frontiers of accumulation, devise fixes for its recurring cycles of crisis of overaccumulation of capital and underconsumption of labour, and deal with the challenge of expanding accumulation while avoiding destruction of its ecological base (O'Connor 1998; Harvey 2003; Rasmussen and Lund 2018). When the forces of commodification gain momentum in a landscape, they transform land and life by reshaping social relations around property (including land access), labour, income, consumption and reproduction, and ecology. These are the five fundamental questions of political economy and political ecology: who owns what, who does what, who gets what, what do they do with the wealth produced, and how do these four questions transform ecology, and vice versa? (Bernstein 2010). During historical moments when land is made to contribute to addressing crises in profit-making, land control grabbing (land as soil, resource and territory) intensifies, and unfolds in and through a landscape to facilitate extractivism (McKay et al. 2021; Andrade 2022; Chagnon et al. 2022). State and social forces that advance continuous commodification of everyday life do so by constantly rejigging the possible combination of structures, institutions and power in particular landscapes and conjunctures, in other words by continuously reshaping land regimes, in the context of the modern state's pursuit of two permanent but contradictory tasks, namely, facilitating capital accumulation while trying to maintain a minimum level of political legitimacy (Fox 1993; O'Connor 2017 [orig. 1973]).

The concept of a land regime offers an important corrective framework to the dominant views of land politics that are either 'too local' (either plot-based or small place-based), or 'too sectoral' (either 'too agricultural' or 'too rural'), which are unable to connect to the wider political economy that determines, directly and indirectly, the causes, conditions and consequences of local or sectoral land politics. Thus, even if the object of specific or tactical interest is a particular locality or sector, it is important to understand these in the broader context of the system. A land regime framework also facilitates thinking about and building broad political coalitions that can withstand the divide-and-rule strategy deployed by social forces opposed to social justice reforms within land politics, such as policies or elites that create cleavages between poor peasants and Indigenous peoples on the issue of land.

Land regime democratization and regeneration

Land regime (see separate entry, this book) is a useful concept for thinking about the politics of land across the board, system-wide, and not just in terms of local plots or sectoral land concerns. There are two important concepts associated with land regime, namely, '*land regime democratization*' and '*land regime regeneration*'.

In academic and public debates, land remains central when discussing the dynamics of social life. The problem is that most scholarship, policy interventions and political advocacies focus on specific elements, or parts, of land issues, and pay less attention to how these connect to broader, system-wide social life. For example, when talking about Indigenous lands, or land issues of urban working classes, the framing is likely to focus on how to demarcate a specific block of land for an Indigenous people's settlement, or housing lots for urban poor. These are of course important, and when done well, could make a positive impact in improving the lives of ordinary people. But land issues pursued this way are liable to get stuck at the level of the 'merely local' or 'too sectoral', and become disconnected from broader system- and society-wide issues.[2] Where this happens, land issues, no matter how progressively they have been tackled, are not sufficient to cause a significant and enduring societal transformation.

Moreover, gains on local or sectoral fronts can be easily cancelled out by losses in another area, or quickly reversed by subsequent social processes. For example, redistributing farmland to small farmers through land reform could be wiped out by the dynamics of commodification credit, whereby small farmers become heavily indebted and have to sell the farm plots they received during the land reform process. Or urban sprawl and the commodification of land that comes with it could cancel out earlier gains from a wetland conservation initiative using mangrove reforestation, as real estate development transforms everything in its way, including the entire wetland, into highly prized real estate lots.

For many people, having formal property claims—whether individual or group—over specific plots of land is important, and is often equated with 'security'. Thus, when we talk about improving the situation in land relations, we most commonly think of policies like land reform, that redistributes land from private large estates to landless or near-landless peasants; ancestral land demarcation in which a landscape or territory is carved out and formally adjudicated to form a settlement or land restitution for those forcibly displaced from their land. However, unless a plot or block of land is considered in relation to a wider land regime, such policies might not contribute significantly to addressing difficult challenges in society in terms of improving the lives and livelihoods of people, of how human beings live on the planet.

Transforming land and social life plot by plot, or local place by local place has benefits, because it deals directly and concretely with actual people inserted into various land-based social relations. But even applying best practices in the field, the overall impact on society will be limited. We therefore argue that a more promising approach to the concrete, system-wide transformation of land and social life builds on the concept of land regime; it is the regime, and not just some elements of it, that should be transformed.

Transforming a land regime requires two conceptual building blocks, namely, processes of democratization and regeneration. We define *land regime democratization* as the process of dismantling land-based social structures, institutions and political power that provide a foundation for exploitation, oppression and ecological plunder that is class-based, racialized and patriarchal, while simultaneously building social justice-oriented land-based social structures and institutions, and recasting the correlation of state and social forces in favour of those who seek to end the unjust system. Its scope is necessarily rural/urban, agricultural/nonagricultural, and its extent will be a matter of degree, spatially and temporally uneven. The correlation of state and social forces for land regime democratization needs to be far broader than for conventional agricultural land reform. Indeed, it will need to be extensive, if not comprehensive, and system-wide. Land regime democratization does not favour a single type of property system; rather, it is based on the assumption of a property regime that is inherently pluralist in character, in which diverse property systems co-exist—a broad-based property regime, a mosaic of private, public and commons.

But while land regime democratization is a necessary condition for land and life transformation, in the era of climate change, it is not sufficient. We can democratize a land regime, but this will not automatically lead to a regenerative use of the land. Historically, far too many partial land reform achievements have later contributed to reinforcing, not healing, the 'metabolic rupture'. *Land regime regeneration* is the process of re-embedding land-based social structures, institutions and political power within socio-ecological relations in ways that do not undermine the ability of ecology to constantly regenerate. Like regime democratization, it is necessarily both rural and urban in geographic scope, inherently uneven across space and time, and is a matter of degree.

The three intertwined dimensions of land regime—structures, institutions and political power—can only be democratized together. Progressively worded laws aimed at democratizing land may be passed, but if there is not sufficient change in the correlation of forces within the state and society, then those laws are unlikely to contribute to democratizing unjust structures. Autonomous social movements struggling for democratic land politics may gain strength, but without a significant shift in formal and informal institutions, it is difficult to expect substantial and concrete reforms in land-based social structures. Land regime democratization and regeneration provide an ethical compass to guide us as we study the dynamics of continuous renewal of life and society in order to shape the character and trajectory of change into something better, more sustainable, fairer and kinder.

Land rights

'Land rights' is perhaps the term most commonly used in association with promoting land tenure security, as in, security is achieved if and when people have land rights. The term land rights is used interchangeably with property rights. While land can be seen as a material object that can be bought and sold, land rights are an issue of social relations. There are two broad types of relational rights: the right to exclude others (private property, state property) and the right not to be excluded (common property) (Macpherson 1978) (see 'Property', this book). Land rights take the form of a 'bundle of rights' and represent 'enforceable claims' (Macpherson 1978), for example, the right to use land, the right not to be excluded from having access to the community forest, the right to own and sell land as a property and so on. The term 'land rights' is commonly used to suggest land tenure security for the traditionally marginalized classes and social groups in society, including poor peasants and Indigenous peoples.

However, having a bundle of rights related to land does not guarantee that the rights-holder will be able to assert control and generate actual benefit from these rights. For example, many peasant households in the Philippines have been awarded farm plots under the land reform programme, but the owners of the expropriated land refuse to recognize the government's redistributive action, and have mobilized their own private security forces, often aided by the courts, to block the land reform beneficiaries from taking control of the land awarded to them by law. There are many of these so-called 'uninstalled land reform beneficiaries' stuck in this situation: they have the bundle of rights, but they do not have the bundle of powers to access, control and derive benefit from the land that, according to the law, is theirs. In another example from the Philippines, a powerful landlord who has his own armed security force and is connected within the local police, military, local government and judiciary, took control of a vast area of public land over which he has no formal rights whatsoever and transformed it into an enormous coconut plantation and cattle ranch. Everyone, inside and outside of government, knows of this occupation by the landlord, but no one dares to question it, and no one demands that the government seize back the land. In this case, the landlord has no bundle of rights, but he has the bundle of powers to take control of the land nevertheless. These two cases illustrate the important distinction between bundle of rights and bundle of powers that Ribot and Peluso (2003) explain (see 'Access', this book).

We are not suggesting that the concept of 'land rights' is not relevant. Quite the contrary. The insights about rights and powers illustrated above contribute to a more critical appreciation and reflection about the concept of land rights. There is nothing inherently good or bad about land rights or about promoting land rights to advance the interest of traditionally marginalized social classes and groups in society. The problem arises when we assume uncritically that land rights are always and everywhere good for and beneficial to the exploited and the oppressed, or if we assume that having land rights provides a guarantee that the rights-holders will be able to take control of the land and derive benefit from it, or that giving no rights to elites disqualifies them, *de jure* and *de facto*, from taking control of some lands.

When working people do not have land rights, the struggle to achieve them can be daunting. Fighting for a national law which mandates expropriatory redistributive land policies is and always has been a formidable prospect. But as Ben Cousins (1997) points out, the struggle to make rights real is just as difficult (see also McKay 2018). In Indonesia; for instance, the Basic Agrarian Law of 1960 was a progressive land policy, but it has basically lain dormant since the mid-1960s.

One of the dangers of the term 'land rights' is that it is often used too loosely and too broadly. First, it is commonly used in the sense of ratifying or formalizing what already exists. In societies characterized by land-based inequality, the term 'land rights' has little to say about purposive redistributive reforms (see '(Re)distributive reform', this book). Second, and related, where laws do exist, the term is used in the sense of technically and legally implementing the terms of the law, which could mean protecting the status quo rather than unsettling it. For example, promoting and protecting land rights in contemporary Colombia means advancing the land rights of *campesinos* but also the land rights of big landed elites. Third, from a 'positive sum' (as opposed to a 'zero sum') position, there is an implicit assumption that such rights can be allocated to some, without others losing their rights to the same land. The concept of land rights usually adheres to multistakeholder principles, which embrace the notion of equal rights for everyone (McKeon 2017): poor peasants, big landlords and agribusiness alike.

Recommended reading

Cousins, B. (1997). How do rights become real? Formal and informal institutions in South Africa's land reform. *IDS Bulletin*, *28*(4), 59–68.

Land rush

From 2008 onwards, the term land rush has often been used interchangeably with other terms such as land grabbing, land deals, land boom and LSLAs. Yet, it has never really been defined for what it is. For us, it important to clarify what the term 'land rush' means, and how is differs from the other terms. Only by knowing what a phenomenon is can we begin to consider what to do about it. What is offered below is a definition of land rush. It has to be understood in conjunction with three closely related concepts, namely, spectacle, land grabbing, land boom (see separate entries, this book).

One problem in the literature and public debates is the tendency to use the different terms interchangeably, when they actually have different and competing connotations. For example, if we refer to the acquisition by Hengfu sugar company of 143,000 hectares of land in Cambodia, mostly from the Indigenous community, as 'land grabbing', this suggests something deeply political and unjust; if we use the term 'LSLA' instead, this suggests a more technical approach, focusing on such aspects as the size of the land and the administrative procedure of acquiring it. The problem around terms also applies at another level, when one term such as LSLA or land grab is used to refer both to the process of acquiring and reallocating a specific block of land or space in a specific moment, on the one hand, and to the generalized land fever and mad *scramble for land* in its amorphous form on the other hand. These are overlapping phenomena, but are quite distinct from one another. The literature and public debates have focused on 'concluded, operational land acquisitions' or land grabs that are usually corporate-driven. But there are other categories of land transactions that are equally important, including 'land rush'. Land rushes, however, have not been seen as a unit of inquiry or subject of public debate, at least not in any systematic way. This has to be corrected, and having a clearer definition of what constitutes a land rush is a step in the right direction.

We define land rush as denoting a chaotic, relatively short-lived, historical juncture marked by a sudden upsurge in demand for land, accompanied by an extremely speculative and competitive, often violent and convulsive transition from one set of rules on commodity and land politics to another. More narrowly, it refers to that insurgent moment when the prevailing 'land regime' (see separate entry, this book) is seriously challenged but not yet fully replaced by a new regime. Land regimes—established patterns of rules on how to govern access, use, control and ownership of land across sectors and in the rural–urban continuum—are thus questioned, reinterpreted, recast, or disregarded amid increasing unruliness. Land rushes simultaneously cause, and are an outcome of, wild speculation and the spectacularization of potential fortunes that might be derived from seizing particular blocks of land or territories, directly or indirectly. Land rush encompasses various elements, namely, land enclosures, land grabs, land deals, land acquisitions, land boom, commodity booms—small, medium and large—and multiple actors (state, non-state, corporate and non-corporate, foreign and domestic investors, land brokers and scammers). A land rush has distinctive sociopolitical features, namely, wild speculations, hyperbolic claims, fantastic spectacles and a convulsive atmosphere, very much along the lines of Anna Tsing's notion of spectacle and spectacular accumulation (Tsing 2000). It is an accumulation and convergence of multiple land grabs in multiple sites linked together in a global contagion, until the phenomenon evolves into a full-scale mad scramble for land, a land rush.

Land control grabbing within a land rush can occur either physically or through a variety of mechanisms, directly or indirectly, especially in the era of financialization and digitalization when effective control of land may well be seized by distant, depersonalized finance capital (Ouma 2016; Salerno 2017; Clapp and Isakson 2018; Fraser 2019; Fairbairn 2020), prompting Ashwood and colleagues to ask 'what owns the land?' (Ashwood et al. 2022). This is all part of a process of 'rendering land investible', in the words of Tania Li (2017), and usually occurs quickly, within a particular historical juncture.

Land rush can emerge from any one—or a combination—of three broad possibilities: (a) there is a sudden massive speculative demand for land but no clear set of rules on how to effectively respond to such a demand; (b) there is a clear existing set of rules on (re)allocating land control and use but this does not allow the surging speculative demand for land to be met; (c) there is a new set of rules on land (re)allocation and land use change and a push by some to get ahead of the rest in taking advantage of such rule changes. In all cases, a land rush has the element of extreme urgency and is relational. The sense of urgency is due to the desire to be ahead of others; a land rush is thus extremely time-sensitive. Its relational character can be seen in at least two ways. First, the actions of any one player can only be calculated in relation to those of others, whether one is offering land or wanting to grab land. Second, the rush is inherently multi-sited geographically, spurred on by wild projections about how much land is needed by land prospectors and the desire of investment prospectors to offer as much land as possible, where comparative features of these offerings can make a huge difference in the outcomes of the land rush.

Here, the geographical spread in land rushes is taken in relative terms: it can be bounded within a national territory, a transnational regional area (Sosa Varrotti and Gras 2021), or it can be global. This geographically multi-sited characteristic also contributes to a temporal unevenness, and the combination of the two means that any major land rush is actually plural and diverse, and is more precisely referred to as 'land rushes' (Messerli et al. 2014). Moreover, we take land control here as a term that has multiple meanings: land control to shape food and agricultural production systems; land control to capture carbon in forestry conservation project; land control to allow for the capture of hydropower in a mega dam project; land control to capture subsoil minerals; and so on. This allows for broader meanings of land control and land grabbing to include 'green grabbing' and 'water grabbing' (Fairhead et al. 2012; Mehta et al. 2012; Ojeda 2012; Franco and Borras 2019; Bruna 2022), land politics as water politics or river politics (Suhardiman et al. 2012; Boelens et al. 2023), or the inseparability of land and water (Cortesi 2021). Holliday (1999) used the term '(world) contagion' in his description of the California gold rushes in the second half of the nineteenth century. Holliday's term has resonance with the recent global land rush as it spread globally in two ways: first, geographically, as it gained ground in the South, North, East and West (Desmarais et al. 2015; Roman-Alcalá 2015; van der Ploeg et al. 2015). And second, in and through spheres of social life, as it reshaped and was shaped by food politics, climate change politics, labour regime and migration, nation-building and citizenship, and geopolitics.

Recommended reading

Borras Jr, S. M., & Franco, J. C. (2024a). Land rush. *The Journal of Peasant Studies*, 1–16. https://doi.org/10.1080/03066150.2024.2317961

Land sovereignty

In 2012, at the height of political debates and academic research interest in global land grabbing, we put together a tentative concept that we thought could help capture the emerging land struggles globally that were not, in our view, fully rendered by the standard framing of land struggles that had been largely confined within the classical framework of 'land reform' or 'agrarian reform'. We proposed the concept of land sovereignty. Since then, our thoughts about it have evolved, particularly towards and around the '5Rs' (see 'Five Rs', this book). But we feel that our original thought processes and building blocks for these ideas, encapsulated in the concept of land sovereignty, remain relevant. Our main intellectual and political agenda here is to inspire productive discussion on how to conceptualize a land pillar within the broader concept of food sovereignty, and the larger struggles of working people more generally.

Our assumption is that the contemporary politics of land have been fiercely contested in ways that are significantly different from land struggles in the twentieth century. At least four late twentieth and early twenty-first century phenomena have shaped land politics, namely, the unprecedented extent of urbanization in the world; the rise of climate change politics (see 'Climate change politics', this book); the large and growing proportion of the working class that are jobless or near-jobless and living with precarious livelihoods in the largely informal sector, estimated to number at least a billion people; and approximately a billion chronically hungry people. Food sovereignty, broadly cast beyond the narrow parameters of food politics, has become a reference point, a rallying cry of various political struggles around these four phenomena. In turn, food sovereignty requires a land pillar. How should it be framed? The mainstream response among agrarian social movements is 'land reform' or 'agrarian reform', and among liberal progressive advocacy groups, 'land tenure security'.

Without devaluing land reform and agrarian reform, we propose a shift in framework to the concept of 'land sovereignty'. Taking seriously the historical demands for land by various strata of working peoples, we need an alternative frame that better expresses a truly pro-working poor class bias in land issues, and especially the core idea of the rural working classes being able to exercise full and effective control over the land where they live and work. While every term—including 'land sovereignty'—has its limitations, we believe 'land sovereignty' comes closest to capturing the essence of the demand for land by rural working peoples.

The term connotes a sense of 'belongingness': the land belongs to the people who work it, care for it and live on it, and, in turn, those people belong to a particular land as a people. It also reminds us that individual and collective plots of land are part of larger socially constructed landscapes, territory and homeland, which in turn reflect the kinds of relationship between human societies and the environment that have emerged over time in a given place. We should always strive for an ecologically healthy relationship with our environment. The issues of space, ecology, territory, identity and belongingness—key dimensions of a truly pro-poor and sustainable conception of and demand for land that can be prioritized and promoted into the future—can be best captured in the notion of land sovereignty.

For us, land sovereignty is the right of working peoples to have effective access to, use of, and control over land and the benefits of its use and occupation, where land is understood as a means for production and social reproduction, and thus, as a productive resource, landscape, territory and homeland (see 'Land in production/social reproduction', this book). Simply put, land sovereignty is the realization of the working peoples' human right to land. Some may question the use of the term 'sovereignty', which might sound awkward. But we believe it can be defended, as we explain below.

As we use it here, 'sovereignty' has a dual meaning. On the one hand, it is a call to action to bring the state back in and hold it accountable to citizens in the face of a global corporate assault on the agrarian front in the form of (trans)national enclosures. This approach must of course address the problems of a state-centric land policy framework. It also requires a deeper understanding of the state through an intersectional lens, taking account of class and interlocking axes of social difference (race, ethnicity, caste, gender, generation, religion, nationality). The key question here is: whose state? On the other hand, 'sovereignty' implies support to working peoples and their human right to have control over land as a productive resource (farmland, grazing land, etc.), landscape, territory and homeland. It is in part a response to the recent popular calls to preserve and protect the commons – not in the context of a commons-only alternative, but with the understanding that a pluralist conception of property, as a mosaic of private, common and state, is probably the more sensible way forward from the current conjuncture. The land sovereignty approach must therefore address the flaws of commons-oriented (populist) frameworks that tend to de-emphasize intra-community class and group differentiation. By taking this mosaic perspective, it may be possible to address some of the key weaknesses of each of the ideal-type property relations (see 'Property', this book).

There are five major reasons why land sovereignty is a useful normative framework. First, in terms of systems of property, it accepts that these can be plural, encompassing private, common and state property, as well as the multiple and diverse hybrid types in between these three ideal types. Land sovereignty does not privilege western private property rights, but nor does it reject them. It sets a premium on preserving and protecting the remaining commons and expanding these, but it does not romanticize the idea, especially since these are sites of intra-community exploitation and oppression. It recognizes the importance of state property as an alternative to full-scale individualized and privatized property rights, but is also cognizant of the contradictory role of the state in terms of contestations around control over land resources and their use, and the dangers of simplification processes.

Second, in terms of public policy options, land sovereignty builds on and privileges redistributive land reform, but then goes beyond it. It looks at land reform as a key policy demand and instrument in places where a western property rights regime has been entrenched, resulting in land concentration and land monopoly. But it is conscious of the limits of land reform. Land sovereignty thus brings into play other equally important redistributive land policies including land restitution, especially where people have been displaced and dispossessed by earlier and current waves of land grabbing. The redistributive character and content of a land policy are not automatically defined and delivered by the particular type or form of a policy: there are redistributive land reforms, but there are also conservative pro-elite land reforms (think of 'market-assisted land reform'; see separate entry, this book), and so on. The content and character of land policies are partly shaped by state–society political contestations (see '(Re)distributive reform', this book). This implies the existence of a variety of land policies whose redistributive content can be shaped and secured through mass struggles: e.g. forest land reallocation policies, community-based forest management, tenure reform, leasehold reform, and so on. The key is to understand the character and direction of the flow of land-based wealth and power transfers.

Third, land sovereignty necessarily takes on board the land-as-a-productive resource focus of land reform, builds on it, and then goes beyond it to include two other master narratives: land as territory and as landscapes. As such, land sovereignty is a framework that does not back away from existing tensions between competing narratives, but rather confronts them. Indigenous peoples' communities and movements worldwide seldom use 'land reform' as their framework of struggle, as they aspire to control their 'territory'. Moreover, land reform has always been a divisive policy frame in particular settings, especially where there are large numbers of rural labourers working either for smallholders themselves or for large farms and Indigenous communities. A land campaign that uses a land reform framework will inevitably alienate at least some of these labourers and Indigenous peoples, many of whom have no interest in becoming peasants and some of whom see land reform as a mechanism for grabbing their ancestral lands. Nevertheless, they may have place-based land demands; for example, where land is cast as a territory, landscape or homeland, labourers may demand not farm plots but house lots. In a similar vein, there are small family farmers who already have their plots of land and so would not see a call for land reform as relevant for them. And yet, they may also have persistent land issues, such as trying to protect their land access for social reproduction (access to common forest, public parks, etc.), which become more visible when land is broadly cast as territory and landscapes. Land sovereignty can thus be a common platform between peasants, labourers and Indigenous peoples and the general working people in rural and urban communities.

Fourth, in terms of positioning mass struggles from below, land sovereignty can provide a more inclusive, global master frame than land reform. Cast broadly, as explained above, land sovereignty can serve as an overarching narrative that can bring together—as fellow claim-makers—social movements from the South and North, from rural and urban settings. This may resolve the problem inherent in an international land reform campaign that is too South-oriented and rural/agricultural-centred. Resolving these divides could unleash a massive global political force worldwide: South–North, rural–urban. Imagine a shared land campaign that brings together peasants from Mali, forest dwellers from Indonesia, peasants from Guatemala, part-time small family farmers in France, Indigenous peoples from Colombia, and urban gardeners in Detroit. This is a broad coalition that is inconceivable within a land reform framework.

Finally, an inclusive global land struggle can be effective only when cast not as a stand-alone issue and campaign, but as one that is able to connect, and connect well, with broader working peoples' campaigns and political projects. Land alone does not constitute an alternative model of development; it is a necessary component of the latter. The current alternative narrative that has inspired tens of thousands of working peoples, middle classes, intellectuals and activists worldwide, both urban- and rural-based, young and old, is food sovereignty. Food sovereignty needs a land pillar, but neither land reform nor agrarian reform has the necessary qualities to be this pillar. Conversely, land struggles need a broader master frame. In this context, land sovereignty has the capacity to connect with food sovereignty, allowing for a mutually reinforcing, synergistic interaction between these two frameworks and the social movements therein. But this also means that—like food sovereignty—land sovereignty is necessarily a cross-class alliance and political project, internalizing both synergy and tension in these alliances and political projects. Moreover, the current global conjuncture has been marked by a recent global land rush, rural and urban, occurring in the South and

North. It is being resisted—albeit weakly, unevenly and on most occasions, defensively—in these various sites of struggle. Land reform or agrarian reform do not provide an adequate master frame for working peoples' struggles against global land enclosures. These enclosures affect multiple social classes and groups in a wide variety of ways. A cross-class, peoples' counter-enclosure campaign has become urgent and necessary, and has been underway unevenly despite lacking an overarching master frame. Land sovereignty can provide that frame.

A *peoples' counter-enclosure campaign* is one in which working peoples resist political processes that maintain an inequitable status quo, or even further (re)concentrate land ownership and control. In the specific current international context, it is a campaign in which people struggle to resist global land grabbing. A pro-active peoples' campaign against land enclosures is just as important as defensive struggles. In other words, a people's counter-enclosure campaign is necessary but not sufficient; what is needed is a *peoples' enclosure campaign*. A peoples' enclosure campaign is one in which working people pro-actively assert their political control over their remaining lands against potential and actual threats of corporate or state enclosure. It can be organized either independently of the central state via community-based enclosure of the commons or in direct engagement with the state. Here, we focus on the latter. There are two types of land policy institutional contexts within which a peoples' enclosure campaign can, and should, take place, namely, 'redistributive' and 'distributive' land policies (see separate entries, this book). There are at least three broad strategies possible in implementing a peoples' enclosure campaign advocating for redistributive and distributive land policies: state-led, social movement or community-led, or state/community-driven. It is not the presence or absence of either the state or community entity that defines these types of strategies; rather, it is the character and the extent of their contribution that is important.

Land sovereignty as a campaign thus represents two politically intertwined struggles of rural working people: the peoples' counter-enclosure campaign (defensive) and the peoples' enclosure campaign (pro-active). The idea of 'land sovereignty' is partly inspired by the relative success of 'food sovereignty' as a framework for an alternative food system, a campaign and a movement. Food sovereignty has become a global framework for all those who are confronted by a 'food question', whether they are in rural or urban areas, in the South or North. It is a concept that is flexibly interpreted and deployed, depending on structural and institutional location, and it has become a truly cross-class alliance political project. This is an inspiration and reference point in our attempt to reframe land struggles, and our exploratory concept of land sovereignty.

Land struggles

In classical agrarian studies, land struggles are almost always understood as struggles over agricultural land, that is, land in the sphere of production.

The broader backdrop is the agrarian question—the location of agriculture in relation to the development of capitalism in a given society. In this context, land struggles are usually seen as struggles to dismantle land monopoly of the parasitic landed classes, in order to redistribute land to landless and near-landless working people. Many of these land struggles were integral components of national liberation and anti-colonial struggles, especially post-World War II. Seen from this perspective, most of the struggles for land were about rural and agricultural land in the Global South. The main social forces pushing for these land struggles were landless and near-landless working people (Moyo and Yeros 2005).

This type of land struggle is still important in many societies today, and remains a prominent part of some national agendas. But it is not as widespread as it was during the twentieth century. In the past two decades, there have only been a handful of iconic struggles of this type, such as those by the Landless Workers' Movement (MST) of Brazil, FENSUAGRO and other militant agrarian movements in Colombia, the early twenty-first century upheaval in Zimbabwe, land reform struggles in the Philippines, Indonesia and South Africa, and a few others. It is tempting to conclude that contemporary land struggles are thin and weak, especially when compared with the land struggle component of the great peasant wars of the twentieth century (Wolf 1969). Yet, if we examine land issues of the twenty-first century using an appropriate analytical lens, then our conclusion will be significantly different.

The changed structure and institutions of the politics of land in contemporary times (see Borras and Franco 2023: Chapter 2) require us to review and rethink the concept of land issues and struggles of working people. One outcome of the transformation of global land politics is that while the relevance of conventional land reform has been reaffirmed, its relative importance has diminished. Contemporary land issues can be clustered into four categories, of which conventional land reform, which might be expressed as *rural/agricultural in the South and North* is only one (type I). In terms of academic research in agrarian studies, this category demonstrates the relevance of political economy perspectives that focus on understanding dynamics of agrarian transformation brought about by the penetration of capitalism into the countryside (Lenin 2004 [orig. 1899]).

The remaining categories (types II, III and IV), which arguably have always existed but were never key themes in agrarian studies, have become more relevant and gained in relative importance. The category *rural/non-agricultural/South and North* (type II) has become, or should become, just as compelling as type I as a category for academic research and political action. It is related to a wide array of climate change mitigation and adaptation initiatives and issues, notably the dramatic expansion of non-agricultural neoliberal conservation and carbon sequestration initiatives; the resurgence of hydropower projects and solar and wind farms; the massive expansion of 'no dwelling zones' in 'fragile areas' due to climatic change; and land in the broader context of social reproduction, including land for home lots, home gardens, and so on (see Shah and Lerche 2020; Borras et al. 2022; Cousins 2024). The sheer number of rural people directly affected by these policies and initiatives, and the logic underpinning them, demand the full incorporation of this category into Critical Agrarian Studies.

The issue of rural–urban, agriculture–industry links (Kay 2009) has remained relevant, and recent developments have rendered it even more central in classical agrarian studies and Critical Agrarian Studies (Pattenden 2023), albeit in a significantly revised way. Recent demographic changes and patterns of capital accumulation have altered some of the traditional urban–rural links and flows including those related to land, labour, dwelling, food, water, forests, the environment and finance. Rural and urban categories have never been so blurred, and the same can be said about the politics around rural–urban linkages. Moreover, and even more importantly, the rural–urban corridor has become the key site for productive and social reproductive activities of many sections of the 'classes of labour' or working people (Bernstein 2006; Shivji 2017; Pattenden 2023).

The categories *urban/agricultural/South and North* (type III) and *urban/non-agricultural/South and North* (type IV) have thus become important in their own right, as capital attempts to seize as many resources, spaces and labour as it can in order to further processes of accumulation, and as working people bring some aspects of agrarian practices into urban spaces (McClintock 2014; Jacobs 2018; Siebert 2020). Since the last decade of the twentieth century, we have seen an explosion of land conflicts worldwide that take place in urban/peri-urban areas, involving both agricultural and non-agricultural issues.

Traditionally, the main antagonism in the countryside, as framed in classical agrarian studies, occurred between peasants and the landowning classes (or the state, representing the landed interest) and principally centred on agriculture and broader issues of national economic development. Today, antagonisms rooted in land are more plural and diverse. Landowning classes, including *latifundistas* and agribusiness plantation owners, remain entrenched and represent key reactionary classes in many societies. But the current context has brought in other social forces that are equally, if not more, vicious. They include new corporate land grabbers, both transnational and domestic; cross-border non-corporate but pervasive individual land buyers such as big farmers, brokers, renters, scammers, swindlers, 'land mafia' (Sud 2014; see also Levien 2021); financial entities including pension funds (Ouma 2016); the use of hyper-modern digital technology in land and food systems (Fraser 2019); an array of non-traditional agricultural investors ranging from auto companies to livestock processors; as well as big-time conservationists (Brockington and Duffy 2011; Arsel and Büscher 2012). Most of the lands being targeted are claimed by the central state, thus transforming the state into a key land broker that enables and facilitates land grabs and often deploys extra-economic coercion.

In settings where the land is needed but the people are not, as framed by Tania Li (2011), villagers are likely to be expelled from their land. This is especially relevant amid the rise of contemporary plantations of various types. But capital is not committed to a particular mechanism or form of land control, as long as a venture generates profit. Thus, land dispossessions may be caused by market relations, alongside a variety of other mechanisms for grabbing control that often involve extra-economic coercion (Levien 2018). For example, rural villagers may not be expelled from the land at all but rather be subsumed in emerging capitalist enterprises.

All of the dynamics described above have revived old and engendered new axes of political conflict, and generated a range of political reactions from below (Hall et al. 2015). When land deals hit the ground, they impact already socially diverse and differentiated communities. They affect different social groups in different ways, provoking political reactions that have multiple, complex and often contradictory dynamics in terms of class and co-constitutive axes of difference: race, ethnicity, caste, gender, generation, religion and nationality (Gyapong 2019).

Recommended reading

Hall, R., Edelman, M., Scoones, I., White, B., & Wolford, W. (2015). Resistance, acquiescence or incorporation? *The Journal of Peasant Studies, 42*(3–4), 467–488.

Moyo, S., & Yeros, P. (Eds.). (2005). *Reclaiming the land: The resurgence of rural movements in Africa, Asia and Latin America*. London: Zed Books.

Land tenure security

Land tenure means 'the rules and arrangements connected with owning land, especially land that is used for farming' (Cambridge Dictionary n.d.). There are multiple and diverse types of land tenure. One is freehold tenure that provides the property owner the full range of rights over a demarcated block of land, including the right to alienate. Others include various forms of commonhold, usufruct tenure which gives one the right to use, leasehold tenure, and so on. We can also differentiate between land tenure on paper and land tenure in reality. The former might comprise neat land tenure categories that are officially recognized, while the latter could be far messier than the neat grid of official categories would project or claim. The main point is that land tenure is a property-centred perspective on the politics of land, very different from the land access-centred perspective that Ribot and Peluso (2003) employ (see, 'Access', this book). The main focus of this entry is security, as in land tenure security.

Land tenure security is the assurance offered to the tenure holder in whatever arrangement applies—freehold, leasehold, and so on—that rights over the land in question will be guaranteed and enforceable by societal norms or state laws. The tenure holder can be a natural person or artificial person; an artificial person can be a corporation or similar which is 'recognized by the state as having the same (or similar) property rights as a natural person' (Macpherson 1978: 5). This means that a land tenure holder can come from the ranks of the working people, or from among the owners of big capital. If the state guarantees such rights formally and officially by law, the corresponding coercive apparatus of the state, namely, police and judiciary, can be mobilized to enforce those rights.

Land tenure securitization is the process that gives land tenure security where this has previously been absent (contrast this concept with 'Land regime democratization', this book). For example, peasants informally squatting on public land without any official agreement with the government can be given more formal and official rights over the plots of the land they occupy and work. Securitization can also refer to providing stronger rights where there have previously been some rights, but these have tended to be weak. For example, where formally recognized tenancy relations between landlords and tenants exist, but landlords can legally expel the tenant anytime without legal consequences, the position of tenants can be made more secure through the provision of the right not to be expelled from the land without due process. The degree of improvement to security, and who is supposed to benefit from this, are not always specified in the use of this concept, as the concept itself tends to be more administrative in nature. This means that it is open to different interpretations. The climate of global land politics could very well be the social force that can influence what land policy terms mean at particular historical junctures (Macpherson 1978: 2–12). In the first three quarters of the twentieth century, land tenure security almost always meant strengthening the position of subaltern tenants (peasants, tenants, sharecroppers). Since the 1980s, in a neoliberal world, land tenure security is more often defined in the context of the market economy. In contemporary usage, land tenure security essentially means ratifying what exists by way of improving the rights of tenure holders over their claimed lands, which often means formal, enforceable rights through state laws. It almost always means promoting private property. The beneficiaries of such an improvement of rights can be poor working people, well-to-do middle class, or owners of big capital.

An extended quote from Klaus Deininger, a senior economist of the World Bank and the Bank's most influential land policy expert, is instructive:

> Clearly defined rights to land and associated natural resources are important for a variety of reasons. First, investments seldom occur on a blank slate. In almost all cases, land and associated natural resources targeted for investment are subject to existing and often overlapping rights held by communities, individuals, the state, or some combination of the three. Understanding and respecting these rights is important if investments are to be socially legitimate and legally secure. Failure to do so can lead to conflict and strife that will negatively affect the economic viability of land-related investments.... Failure to map and record land rights, even if only at the community level, makes it difficult to identify boundaries and legitimate owners as a basis for engaging in mutually agreed-to land transfers. Recording rights provides outside

> investors with somebody to talk to, a legitimate and authorized partner to negotiate the nature of investments and compensation. A formal record is also very much in investors' interest as it reduces the scope for fraudulent transactions and the need for costly inquiry to prevent the surfacing of possible undisclosed prior claims or overriding interests (e.g. land use restrictions).
>
> (Deininger 2011: 236)

There are instances when Deininger also talks about the benefit of improving land tenure security to small farmers. But it is clear to us that the main agenda of new institutional economists like Deininger, and international financial institutions like the World Bank, is to offer security to the owners of capital, the investors, to encourage more capitalist investments. Land tenure security in this context is essentially the formalization and facilitation of further commodification of land and nature.

Land tenure security is often used interchangeably with the concept of land reform. They are not the same, and should not be conflated. Land reform in its classic definition means redistributing lands from large private estate to landless and near-landless peasants. Land tenure security does not necessarily call for redistribution of land.

Land tenure security is neither inherently good nor bad as a concept in the context of advancing social justice. It can be interpreted to explicitly benefit working people. But such an interpretation is political, and depends on prevailing correlations of state and social forces in a given society and conjuncture. In the current neoliberal world, however, when land tenure security as a term is not explicitly defined in a particular context and time, then its default interpretation is likely to be what the dominant ideological forces define it to be, such as Deininger's interpretation. There is therefore a danger that in important (inter)governmental documents, such as IPCC's regular reports in which land tenure security is often and prominently identified as a key policy instrument necessary in addressing climate change, this is translated in reality to mean formalizing instruments for further commodification of land and nature, and securing investments of owners of big capital, and not necessarily land tenure security for working people.

Recommended reading

German, L. (2022). *Power/knowledge/land: Contested ontologies of land and its governance in Africa*. Ann Arbor: University of Michigan Press.

Landed class

In the study of the politics of land in classical agrarian studies, one central concept is land-based class relations. The defining feature of class in agrarian political economy is the social relations of the means of production from which people in the same category (e.g. landless, or smallholcer) tend to pursue similar ways in building a livelihood and/or accumulating wealth, and in doing so, interact with other social classes and groups under different terms, e.g. relations with moneylenders, merchant capital, etc. (Herring and Agarwala 2006; Bernstein 2010). Class is about relations, and is thus inherently dynamic and historical (see 'Class' this book). This is different from other concepts of class based on spatial categories, as in Lipton's notion of rural and urban classes, for example (see Byres 1979).

Following this basic definition, the concept of 'landed class' as used in Critical Agrarian Studies refers to social groups that have control of the means of production, that is, land, and by virtue of this, are able to generate, appropriate and accumulate income and wealth—in other words, a dominant group that owns vast lands and accumulates wealth by extracting ground rent from those who work the land, either through land rental or sharecropping. This is a typical example of feudal/semifeudal landlords who generally do not invest in modernizing production systems or in technological advancement, but are content with extracting wealth from peasants via land rent or crop shares. This kind of landed class was dominant in many developing countries in the twentieth century, and held the monopoly control of agricultural lands in many societies. They are loosely referred to as traditional or backward agrarian landlords. They were the most common target of redistributive land reforms of capitalist development orientation during the twentieth century, on the basis that such landholdings act as barriers to achieving modern capitalist economies and/or violate the social function of property.

There are other socio-economically dominant classes that have control over and interest in land, but have a different class position in society more generally. Some landed elites with a land monopoly managed to modernize, for instance through technological advancement and by transforming relations with peasants into capitalist forms through wage relations or contract farming. Their main wealth is not derived from ground rent, but from appropriating value generated in producing commodities through the use of wage labour and technology, or through various contract farming arrangements with independent peasants. This is the agribusiness type of landed class, and their class interest is in technological advancement for productivity increases which they then appropriate.

Other dominant classes, such as industrial, merchant and financial capitalists, do not always have a direct and significant interest in agricultural land. On the other hand, given rapid rates of urbanization in the world, the rise of real estate capitalists as a class has been phenomenal; in many cases, they combine the basis of their wealth generation across sectors of real estate, banking and finance, trading (especially retail trade) and industries. Financial capital (pension funds, banks, etc.) has emerged as one of the fastest growing 'landed classes'. For example, the Teachers Insurance and Annuity Association of America (TIAA), with its investments in land in the USA, Brazil and other countries, is now the biggest agricultural investor in the world.

Below these big owners of capital are rich farmers. Although their landholdings are not as vast as some of the landed classes above, they are bigger than those of average farmers. The economic power of rich farmers often derives from agricultural productivity increases, as well as from businesses in trading and moneylending. Historically, the conventional, ground rent-dependent landowning class and rich farmers were the two groups most opposed to redistributive land reforms. In capitalist development-oriented land reforms, highly productive modern agribusiness has usually been exempt from land redistribution. There have been occasions when other sections of the dominant classes—those that do not have a direct and significant interest in agricultural land—have even sided with the social forces that were pushing for redistributive land reforms. In some societies, religious institutions have acquired vast tracts of lands, and can be categorized as part of the landed classes, such as the so-called 'friar lands' in the Philippines. In other cases, the military has grown so politically and economically strong that they have acquired interest in and have appropriated vast lands, such as in Myanmar.

Conservation organizations have now been around for several decades, and some have engaged in enclosing lands using a variety of what is called 'fortress conservation', whereby local communities are prohibited, fully or partially, from accessing demarcated spaces, often forests and other biodiversity areas. The power of the big conservation organizations, and the areas they have enclosed and are planning to enclose, have expanded exponentially in the era of climate change politics, with conservation enclosures being justified in the name of climate change mitigation and adaptation as well as biodiversity conservation.

Land monopoly by landed classes and big conservation organizations has persisted and even expanded in many societies today. But there have been qualitative changes in the class and institutional interests which underlie the monopoly of land by these dominant classes and groups. Conventional arguments on the basis of either 'social function of property' (see separate entry, this book) or economic development may now be only partially effective and relevant. Justification for redistributive land reforms has to be rethought, reframed and adjusted to twenty-first century conditions.

Landscape

In agrarian political economy, the dominant view on land comes from the perspective of production, that is, land as agricultural land, while the general treatment of land is in the context of the agrarian question that problematizes the location of agriculture in the uneven development of capitalism. There are two complementary units of analysis and objects of policy intervention and political struggles, namely, demarcated farm plots and the overall agricultural sector. Land reform, land restitution, land tenure improvements via leasehold reform, formalization of rights and land titling programmes often use the demarcated farm plot as the unit of intervention. Research on the success or failure of these policies then focuses on farm plots and their aggregated numbers. We can take, as a hypothetical example, a record of the redistribution of 1,000 farm plots, amounting to 5,000 hectares of land, to 1,000 peasant households at an average of 5 hectares per household. These are often aggregated sectorally. Again, hypothetically, this might mean that 50% of all agricultural landholdings were expropriated and redistributed to 40% of all landless and near-landless peasants. This is the classic treatment of land in agrarian political economy, historically and currently. The strength of this perspective is that it helps disaggregate political economy issues of agrarian relations and their location in national economic development to the household and landholding levels, giving more concrete insights into the social relations between landless and near-landless peasants and landed classes, and the role of the state in reproducing and maintaining, or unsettling, agrarian social structures.

But this kind of classical treatment of land misses important elements of political economy. Land as productive resource is important, but limited, centring our attention on land in the production sphere. If we take the 'land in production/social reproduction' connected whole (see separate entry, this book), then our inquiry will necessarily have to go beyond land as farm plots and the like. We will then have to assign equal importance to land in social reproduction, which entails access to a plural and diverse range of land and nature, such as access to community forest for foraging and gathering of non-timber forest products, access to a riverbank to fetch water and do laundry, access to public spaces for children's playing areas, and so on. If we plot these layers of access to land for production and social reproduction, then we will see a disorderly web within a much broader space and more complex social relations. We will also realize that access to a farm plot is just one of the many forms of land access that are required for the incessant renewal of life and community of rural dwellers. It is the integrity of these webs of access, and not just having access to a farm plot, that enables renewal of life. When a broader lens is used in this way, then 'land as farm plot' in the context of the countryside becomes insufficient. What is important is land in production and social reproduction—the cumulative social relations enacted through the multiple layers of land access across three ideal types of land property which co-exist as a mosaic: private, commons and state property. This also implies the existence of social structures and institutions that enable, threaten or unsettle such a web of land access within a given space and time. Once we centre our attention on this web of land access, the limits of a farm plot perspective and sectoral approaches become clear. At a time when social sciences have moved towards a better understanding of land in production and social reproduction, as well as the relevance of human–non-human relations, the concept of 'landscape' becomes increasingly relevant to the study of land in Critical Agrarian Studies.

In the context of Critical Agrarian Studies, a landscape is a space in which a minimum level of social structures and institutions exist that shape the land in production and social reproduction. It could be a plantation, or a plantation in relation to adjacent non-plantation spaces and communities, or a space with multiple plantations, a rural/urban and agricultural/non-agricultural space and communities, and so on. It could be contained within an administrative unit of a region, or straddle administrative units. Ultimately, the scale of the landscape will vary depending on what a particular inquiry, policy process or political action is interested in.

From the perspective of political economy and geography, landscape is 'a produced, lived, and represented space constructed out of the struggles, compromises, and temporarily settled relations of competing and cooperating social actors: it is both a thing... and a social process, at once solidly material and ever changing' (Mitchell 1996: 30). Mitchell elaborates:

> one of the purposes of landscape is to make a scene appear unworked, to make it appear fully natural. So, landscape is both a work and an erasure of work. It is therefore a social relation of labor, even as it is something that is labored over. To ignore the work that makes landscape, it seems to me, is thus to ignore a lot of what landscape *is*... that people do of necessity. Marx refers to the socially organized labor of people as a 'metabolism' between humans and nature, and the metaphor is exact: the work of people (re)produces a (socialized) nature – and, perforce, it produces landscapes, it transforms the land.
>
> (Mitchell 1996: 6)

The importance of the concept of landscape in land politics becomes immediately obvious once we understood land as a land in production/social reproduction connected whole, across a private, common and state property mosaic, as only a landscape can enable us to conceptualize and empirically observe such a definition of land.

Another fundamental concept in agrarian political economy is 'social differentiation among the peasantry' (Lenin 2004 [orig. 1899]). A key element in understanding this concept is land property relations. Again, land is often understood primarily as a productive resource, that is, agricultural land. Land sales and rental, or land grabbing with the use of extra-economic coercion, are mostly directed towards land in production. When we shift our understanding of land to land in unitary production/social reproduction, it becomes necessary to de-centre our understanding of land accumulation and dispossession away from the sphere of land in production only. Once we do this, then the notions of accumulation and dispossession can be complemented by the concept of 'diminution' of land access, that is, partial dispossession or partial decrease or degradation of the range of access (see 'Diminution', this book). It is also important to think about notions of 'landscape degradation' (see 'Degradation', this book), or indeed, 'landscape reform'. Finally, when we take landscape as the unit of analysis and the object of policy intervention and political action, it becomes increasingly necessary to broaden our definition of social relations to include human–non-human relations, something that logically fits with our broader understanding of social reproduction that includes socio-ecological reproduction (see 'land in production/social reproduction', this book) (Tsing et al. 2019).

Recommended reading

Mitchell, D. (1996). *The lie of the land: Migrant workers and the California landscape*. Minneapolis: University of Minnesota Press.

Tsing, A. L., Mathews, A. S., & Bubandt, N. (2019). Patchy Anthropocene: Landscape structure, multispecies history, and the retooling of anthropology: An introduction to supplement 20. *Current Anthropology*, *60*(S20), S186–S197.

Large-scale land acquisition

LSLA is one of the terms associated with the early twenty-first century global land rush. It is popularly known by its abbreviation, LSLA. It is often used interchangeably with other terms, such as large-scale land investments (LSLI), and large-scale land transactions (LSLT).

The focus of LSLA is on *scale*, which is often interpreted as the size of land, in hectarage, being acquired and reallocated. The cut-off size to qualify for this category has varied over time, and depending on the groups defining it. In the very early phase of the public discussion, the United Nations Food and Agriculture Organization (FAO) suggested 10,000 hectares or bigger as the defining size. This was quickly eclipsed by a much smaller benchmark of 1,000 hectares. The most common minimum sizes used now are 500 hectares and 200 hectares. The term also highlights the *administrative* character of this form of transaction, that is, the act of acquiring the rights over land from one party and reallocating them to another party, often assumed to be a corporate entity.

The most influential organizations that framed, institutionalized and popularized the use of LSLA as a term are the World Bank and the International Land Coalition (ILC). ILC is a coalition of international financial institutions like the World Bank, UN organizations, donor agencies, development NGOs, and some farmers' organizations. In 2012, ILC, together with other groups, established the Land Matrix, the largest data-banking initiative focusing on LSLAs. On its website, the Land Matrix declares itself to be: 'promoting transparency and accountability in land acquisitions.... an independent monitoring initiative that promotes transparency and accountability in decisions over LSLAs in low- and middle-income countries by capturing and sharing data about these deals' (Land Matrix Database n.d.). By highlighting its effort to promote transparency and accountability in LSLAs, it emphasizes the importance of *procedural* matters.

Besides the focus on the size of the land, the rights being transferred and the transparency issue, earlier definitions of LSLA included the question of whether LSLA violated human rights of people who have competing claims over the contested land. There is nothing wrong with the term LSLA and its fundamental emphasis on the three dimensions—size of the land, rights being transferred, degree of transparency—if these are the issues we are interested in or want to study. However, if we apply a political economy lens, we then have to re-orient our inquiry to the causes, conditions and consequences of LSLAs. This will bring us squarely to commodification processes for profit-making as a central driving force in the global land rush, where land is at centre stage for both production and social reproduction under global capitalism. It is important therefore to look into the omissions, silences and contradictions that the term LSLA entails.

First, to be eligible for the databank, an LSLA case should ideally be a demarcated plot of land, around which a legal-administrative contention revolves between the prior claimants of the land, and those facilitating the acquisition and reallocation (usually the state) and their beneficiary (usually a company). This unit of inquiry and intervention inadvertently takes one's focus away from the 'land rush' (see separate entry, this book) which is a global system-oriented approach to understanding the causes, conditions and consequences of the land rush in relation to the contemporary dynamics of global capitalism—a wider task that a purely LSLA-based study may only partially accomplish.

Second, there are three currents that can be generated by a land rush, namely, operational land deals, non-operational and failed land deals, and pin prick land accumulation (see separate entries, this book). LSLA-based data-banking is able to capture the first two categories to a significant extent, but is inherently unable to capture the full scope and character of pin prick land accumulation.

Third, the tendency of the LSLA approach to highlight the scale or size of land involved in the transaction is important, as long as it does not lead to what Marc Edelman calls 'the fetishization of the hectare' (Edelman 2013: 488). One of the dangers of over-emphasizing land size is losing sight of the 'scale of capital'—how and at what scale capital operates, and with what impact on social relations that are directly or indirectly linked to the contested land.

Finally, the Land Matrix website highlights the focus on LSLA in low- and medium-income countries, where it is implicitly assumed that degrees of governance transparency and accountability are low. The flipside of this is the implicit suggestion that LSLAs do not occur in high-income countries, or when they do, they are not something we should worry about; this contradicts several studies that show otherwise (e.g. Desmarais et al. 2015; van der Ploeg et al. 2015). An emphasis on procedures leads to important questions, such as whether or not the transaction is transparent and accountable, or how land acquisition can be made more transparent and accountable. At the same time, this generates more difficult, even if polemical, questions, such as 'which is better: transparent land grabbing or non-transparent land grabbing'?

The principal reason for the popularity of the term LSLA in public debates and scholarly research is that the concept allows large databanks, such as the Land Matrix, to operate, given that demarcation in spatial, temporal and legal-administrative terms is key to quantifiable and bankable data production. It is an important and extremely useful concept that has contributed to a better understanding of the scope, character and trajectory of the early twenty-first century global land rush. But one has to be conscious of the blind spots that are inherent in this concept, as discussed briefly here. This does not make the term LSLA wrong or problematic; it simply reminds us of the analytical limits of the term. It is best used in relation to other parallel terms, such as 'land grabbing', 'land rush', and 'pin prick land grabs' (see separate entries, this book) that are distinct but overlapping.

Notes

1 This entry draws on Borras (2007).
2 Our concept of 'merely local' and 'too sectoral' is inspired by and builds on Fraser (2021) in her critique of environmental justice struggles as 'merely environmental'.

Recommended reading

Cambridge Dictionary. (n.d.) https://dictionary.cambridge.org/dictionary/english/landtenure.
Land Matrix Database. https://landmatrix.org. Accessed 26 June 2024.

M

Market-assisted land reform

Market-assisted land reform—also called market-led agrarian reform (MLAR), market-based land reform, neoliberal land reform, negotiated land reform, and 'willing seller/willing buyer' land reform—is a voluntary transaction between willing sellers and willing buyers. It was framed by mainstream economists coming from the neoclassical economics/new institutional economics tradition, especially those working in the World Bank, as the land reform component of neoliberal economic reforms more broadly. It aims to replace the classic, conventional redistributive land reform which is based on compulsory expropriation. The first nation-wide adoption of this framework was in Zimbabwe, where it was incorporated in the Lancaster House Agreement in 1980, at the end of the Rhodesian war. Since then, it has been attempted in many countries including Colombia, South Africa, Brazil, and the Philippines, driven and funded by the World Bank. In 2024, it is still the main 'land reform' framework being promoted by the World Bank. A slightly modified version of it also remains the main land reform framework in Colombia, based on voluntary buy-and-sell land transaction. So, what is market-assisted land reform? The discussion below revisits the classic, near-textbook formulation of this land policy framework.

Buying and selling of land

It is important to start with the critique of classic, expropriatory land reform by neoclassical economists/new institutional economists. Deininger and Binswanger (1999: 267) capture the essence of the pro-market critique of state-led approaches to agrarian reform when they conclude: 'most land reforms have relied on expropriation and have been more successful in creating bureaucratic behemoths and in colonizing frontiers than in redistributing land from large to small farmers'. The pro-market critique is the most unsympathetic, but arguably the most systematic, critique of state-led approaches to agrarian reform from a purely free market perspective. From this critique, the MLAR model was constructed. Deininger (1999: 651) explains that the

DOI: 10.4324/9781032658438-9

MLAR model is a 'mechanism to provide an efficiency- and equity-enhancing redistribution of assets'. Deininger and Binswanger (1999: 249) claim that 'this approach can help overcome long-standing problems of asset distribution and social exclusion'.

According to the proponents of MLAR, cooperation of landlords is the most important factor for any successful implementation of land reform. This is the MLAR's guiding principle. Hence, it is a voluntary programme: only the land of landowners who voluntarily sell will be touched; landowners who do not want to sell will not be compelled. Deininger and Binswanger (1999: 267) note that 'this approach... aims to replace the confrontational atmosphere that has characterized land reforms'. The willing sellers are paid 100% spot cash based on the full market value of their lands. Deininger (1999: 663) claims that this provides 'a strong incentive for landowners... to sell land'.

Deininger and Binswanger (1999) claim that the creation and development of efficient and competitive individual family farms is the main objective of the MLAR project. The MLAR model adopts a 'demand-driven' approach in land and beneficiary targeting: only poor families who explicitly seek land and only the lands that are in demand by potential buyers are negotiated for the reform programme (qualifying beneficiaries will be provided with funds to enable them to buy lands). In order to strengthen the bargaining power of the buyers during the land purchase negotiation, beneficiaries have to form an organization. The formation of a beneficiary organization is also necessary to achieve economies of scale in the input and output markets, and to carry out a 'peer monitoring' process in order to bring down the programme's transaction costs. The 'fittest' beneficiaries are found through a 'self-selection' process among the prospective buyers. This is expected to exclude less-promising applicants because their peers would not allow them to join the organization that negotiates for the land purchase and credit access.

The model adopts a decentralized method of implementation on the assumption that this will lead to speedy transactions and to transparency and accountability. 'It privatizes and thereby decentralizes the essential process [of land reform]', explains Binswanger (1996: 155). It is partly in this context that MLAR needs local government agencies—for land purchase mediation and tax collection. Local government is assumed to be nearer to the people and so should be more responsive to the actual needs of local communities. According to Deininger and Binswanger (1999: 267–268), 'the MLAR promises to overcome some of the informational imperfections that have plagued the implementation of land reform by government bureaucracies', via localized market information systems provided by local government units. In addition, the MLAR model is assumed to be faster because, as Binswanger (1996: 145) suggests, it 'avoids years of delays associated with disputes about compensation levels'. Moreover, land prices are expected to be lower because of the 100% cash payment to landlords who can factor away transaction costs incurred under the state-led cash-bonds staggered mode of payment.

Meanwhile, 'in a clear departure from the traditional approach, the new model would stimulate, rather than undermine, land markets' (Deininger and Binswanger 1999: 267). Prohibitions on land sales and rentals should be abolished to allow for a more fluid land market (Deininger and Binswanger 1999: 269). MLAR proponents assert that 'closing the gap between agricultural land values and market values of the land makes land more affordable and enhances repayment ability because buyers of land will now find it easier to repay a loan from the productive capacity of the land itself' (van Schalkwyk and van Zyl 1996: 333). This can be done partly through subsidy withdrawal (from large farmers), progressive land taxation, systematic land titling, land sales and rental liberalization, and better market information systems. The MLAR model has a better chance of success if there is an efficient land-titling system. Bryant notes that, '[a] "willing-buyers, willing-sellers" formal land market requires that the sellers can certify that boundaries have been demarcated and that the land in question is legally owned by the seller.

Postland purchase farm development

The programme implementation sequence of the MLAR model is 'farm plans before land purchase'; the argument is that farm development is therefore assured because no land will be purchased without viable farm plans that emphasize diversified, commercial farming. Because beneficiaries are given a cash grant to develop their farms, results will be quickly achieved (Deininger 1999: 666). A portion of this grant must be spent on privatized, decentralized extension services that are strictly demand-driven. Beneficiaries can hire consultants (e.g. NGOs and cooperatives) to assist them with project plans—an approach that is seen by Deininger (1999) as efficient since accountability between beneficiaries and service providers is direct and the process transparent. Widespread credit and investments are expected to come in quickly because land is acquired via outright purchase and land titles are honoured as collateral for bank loans. Moreover, obstacles to investment, such as prohibition on land sales and rentals by beneficiaries, do not apply in the MLAR model, making it more attractive to investors (Deininger and Binswanger 1999: 265).

Financing

The MLAR model adopts a flexible loan-grant financing scheme. Each beneficiary is given a fixed sum of money. The beneficiary is free to use the fund, but whatever portion is used to buy land is considered as a loan and has to be repaid by the beneficiary (100% of the amount at market rates on loan interest rates). Whatever is left after land purchase is given to the beneficiary as a grant to be used for post-land transfer development projects and does not have to be repaid by the beneficiary. This flexible approach is a safeguard against possible fund manipulation and instils the value of 'cosharing' of risks to avoid a hand-out mentality among beneficiaries (Deininger 1999). It also departs from universal subsidies, as it is argued that grants 'are superior to subsidies because they are immediate, transparent, can be targeted and their distortive effects are small' (van Zyl and Binswanger 1996: 419). This mechanism is also thought to be key to reducing the cost of land because it is assumed that peasants will seek the best bargain for their money (Deininger 1999). Finally, proponents of this model claim that MLAR is much cheaper than state-led land reforms primarily because: (i) it does away with huge, expensive government bureaucracies; (ii) land prices are assumed to be lower, and (iii) beneficiaries shoulder 100% of the land cost. The model requires national governments to bankroll the initial phase of the programme, but in the long term, it counts on private banks for the primary financing of the project.

The above is the textbook description of the MLAR model. As mentioned earlier, in countries where this has been adopted, the framework has been adjusted in various ways. However, the fundamental elements described here remain constant, including its voluntary nature and the willing seller/willing buyer foundation of the approach. The outcomes of its adaptation and implementation in many countries have been uneven, but generally dismal, especially when compared to the record of conventional social justice-oriented land reforms. But as late as 2024, the World Bank and several countries continue to promote this framework. Basic elements of MLAR are also deployed in other land policies related to climate change mitigation and adaptation, like REDD+ and climate-smart agriculture, including its demand-driven, voluntary nature, the willing seller/willing buyer principle, the self-selection process, its economic efficiency in monetary terms and so on.

Recommended reading

Lahiff, E., Borras Jr, S. M., & Kay, C. (2007). Market-led agrarian reform: Policies, performance and prospects. *Third World Quarterly, 28*(8), 1417–1436.

Minimum access/size ceiling

Redistributive land policies that aim to establish a more or less egalitarian distribution of land access and control, whether for socioeconomic or political reasons, are not possible without two minimum elements, namely, 'minimum access' and 'size ceiling'. Many of the conventional state-driven land reforms of the twentieth century, especially those that aspire to the iconic principles of 'land to the tiller' or 'land for those who work it', included these two dimensions, sometimes referred to as the 'heart and soul' of land reform.

The idea is straightforward: to democratize access, all those who need land in order to construct a livelihood should have their own land. But this is clearly not compatible with the right of anyone who owns land to keep as much land as they want or the possibility for those who can afford it to accumulate as much land as they choose. Land supply is finite, while the appetite of the owners of land and capital and other political elites to accumulate as much land as they can is voracious. Thus, land to the tiller literally means that if one does not work the land directly then one has no right over that land. This is an ethical position, and land reform laws in this tradition translate it into statutory rules. Historically and across societies where these principles have been adopted, the size of the minimum access (the 'floor') and the maximum access (the 'ceiling') has of course varied, depending on several factors such as geographic location, crop and commodity production and agronomic conditions, mostly defined in the context of land for production (for further discussion on how this differs from 'land in production/social reproduction', see separate entry, this book).

Minimum access is generally defined as the size of a farm (measured in land area) that offers a viable livelihood for a household. This concept has had different labels in different countries at different times in history. A generic term used in the English language is 'economically viable farm size'. This is the term used most commonly in land reform language in the Philippines, where an economically viable farm size would be roughly 1 hectare for irrigated paddy field, 3 hectares for rainfed rice field, 2–3 hectares for upland coconut, 5 hectares for agro-forestry, 1 hectare for commercial banana and 1 hectare for sugarcane. These are reference points to guide the size of land redistribution rather than strict legal rules. For example, the sugarcane estate of former Philippine President Corazon Cojuangco Aquino was redistributed to all plantation farmworkers, but only after the plantation owners had managed to carve out a substantial area of land from the land reform process. This resulted in a much smaller area of land being redistributed to the eligible land reform beneficiaries; in the end most beneficiaries got only about 0.6 hectare, which is not economically viable for farming sugarcane. No wonder that after just a few years, nearly two-thirds of the land reform beneficiaries sold their land.[1]

In Colombia, the notion of an economically viable farm size is represented by the abbreviation UAF (*Unidades Agrícolas Familiares*; Family Agricultural Units), as incorporated in Law 160 of 1994 (Article 38). The size of UAF varies. For instance, in some parts of the savannas of Altillanura, it can be between 500 and 800 hectares; this differs in other parts of the country. Households who were given land in Altillanura to establish *fincas* got similar size of plots through the internal colonization programme, through the distribution of so-called 'public lands' or *baldios*, many of which were also claimed by Indigenous communities. Many of these *finca* owners would later sell their lands in terms of UAF units. So, a normal conversation in land markets in this region of Colombia might go like: 'how many UAFs did you buy'? Or, 'the 43,000 hectares bought by this company, how many UAFs was that in total?'.

The notion of size ceiling means that no one should be able to buy more land than a fixed maximum. This can be by sector or by geographic area or across the board. Sectorally, a government facing mounting protests from the public about perceived land grabbing in their country may be forced to set a maximum size, limiting what a company can acquire for land investment or land concession. In the early phase of the global land rush in the early twenty-first century, there were no such limits on companies: the Daewoo car company acquired 1.3 million hectares in Madagascar, Pheapimex got 300,000 hectares in Cambodia and Karuturi originally got 300,000 hectares in Ethiopia. Widespread protests then forced these governments to introduce a cap. The Cambodian government declared that no company could get more than 10,000 hectares of land for a concession. However, the rules were easy to evade: an influential businessman in Cambodia established two companies, which had the same office address, took two 10,000-hectare blocks of land adjacent to each other, in the central part of the country, and operated the two plots as a sugarcane plantation with one central mill. In another case in the northern part of the country, a large company from China acquired 43,000 hectares of land through four concession contracts, circumventing the 10,000-hectare ceiling rule.

Most of the generation of redistributive land reforms in the twentieth century had much lower land size ceilings and were usually applied across the board and not sectorally. The 1988 land reform law in the Philippines declared that all land, regardless of crops, size and productivity level, would be redistributed to all eligible poor peasants and landless agricultural workers and that owners of the land would be entitled to retain only 5 hectares of land (plus 3 hectares for each adult child of the landowner who was willing to work the land). Of course, owners of big estates mobilized and used their power to circumvent the law by subdividing their estates among dummy owners, etc., and a significant number managed to avoid expropriation this way. But many were not able to manipulate the law and had their lands expropriated. Seen from a different perspective, essentially, the law was saying that no one should own more than 5 hectares of agricultural land regardless of crop and productivity condition. This is an example of a progressive land reform with a land size ceiling rule.

Economists from new institutional economics see this as state coercion, which is assumed to be bad for the free market. It is not surprising therefore that among the first development policies that were dismantled when neoliberalism kicked in in the second half of the 1970s was land reform, starting in Chile with the military regime of Pinochet reversing the earlier Allende land reform. It is not that new institutional economists do not believe in land reform, because they do; but for them, land reform must be a reallocation of land from economically inefficient land use and land users and should be done in a 'demand-driven' (demand by those who want land) and voluntary way, and through economic incentives/disincentives.

These two elements of conventional land reform run counter to the logic of market-based land policy: the principle of 'minimum access' because mainstream economists see the process as 'supply driven' and as such assume it will include incompetent and inefficient land reform beneficiaries, and 'land size ceiling' because mainstream economists assume it will take land away from efficient producers. These two elements are therefore the principal targets for expunction by mainstream economists. No wonder, then, that we see so few land policies in the world today that still have these two components, at least formally; the Comprehensive Agrarian Reform Law of the Philippines is one of only a handful of exceptions. But if our starting point in a land policy is social justice, then there is no way we can avoid the twin principles of minimum access and land size ceiling. Without a land size ceiling, even if there is a formal call for minimum access, it is unlikely to become a reality because there is simply not enough land to be redistributed to poor peasants, landless farmworkers and Indigenous peoples, as most of the land will already have been monopolized by those who can afford it or those who have political power to accumulate land.

It is useful to read this concept of 'minimum access/land size ceiling' in comparative perspective with the concept of 'social function of property' (see separate entry, this book).

Note

1 From a household survey we carried out in 2022 in two villages of the land reform site.

N

Non-operational (and failed) land deals

The term 'non-operational land deals' (or 'non-operational land acquisitions'), often used interchangeably with the term 'failed' land deals, is one of the concepts to emerge from the huge literature on global land grabbing that has accumulated since around 2008. It should be read together with its flipside: 'operational land deals' (see separate entry, this book). This concept is more complex than it first appears, and understanding its nuances will help sharpen our grasp of the politics of land.

This concept emerged primarily because many publicly announced land investment projects, especially those publicized in grand spectacular promotional events or in government and corporate press releases about mega land investment projects, did not progress to any formal arrangements. Moreover, many large-scale land investments, especially those that made extravagant claims about what would come out of these planned capitalist enterprises—from 'plenty of jobs' to 'helping solve the food and fuel crises of the world'—later stalled and were cancelled or abandoned. Many observers thought that most of the hyperbolic claims about large-scale land investment were just that: hype and hyperbole (Kaag and Zoomers 2014).

For this reason, and partly in response to earlier criticisms of the Land Matrix database and its methodology (Edelman 2013; Oya 2013; Scoones et al. 2013), the Land Matrix has clarified the nuanced categories of its databank, which are based on the status of land deals. Thus, the concept of 'non-operational land deals' has become increasingly commonly used in the literature to differentiate it from 'operational land deals'. However, different interpretations of the term 'non-operational' abound, and these differences of interpretation are not technical or administrative in nature, but conceptual.

The Land Matrix uses four categories to describe the status of land acquisitions, some of which contain sub-categories. These are: (1) 'Concluded', which means that a contract was formalized, with sub-categories: (a) not yet started, (b) startup phase, (c) in operation, (d) abandoned, and (e) 'none', which signifies no information is available; (2) 'Failed' means the deal was started but abandoned; (3) 'Intended' means that the deal has neither been

DOI: 10.4324/9781032658438-10

concluded nor failed; and (4) 'Other', which covers two sub-categories: (a) contract expired and (b) change of ownership. The most common interpretation of non-operational land deals includes categories (1d), concluded but abandoned and (2) failed: deals were started but later abandoned. This seems quite straightforward. However, while such technical/administrative categories are useful for certain types of research and policy processes, they have important limitations, especially when inquiring into the political economy of land deals. This can be seen in a number of ways.

First, within the parameters of the Land Matrix dataset and methodology, and the dominant interpretation of such datasets, it is not always clear what is being referred to as 'non-operational' and 'failed': is it the land deal that is non-operational or failed, or is it the promised or planned capitalist enterprises? These two do overlap, but they are distinct from each other. This is an important issue because the labels 'non-operational' or even 'failed' have been applied to cases such as the 28,000-hectare land of the BHO company in Gambella, Ethiopia, in which, after a few years of cultivation, the company suddenly abandoned the area. But the land did not revert back to the villagers; instead, the Ethiopian Development Bank (EDB) took control of the land because BHO had taken a loan from the bank. There are far too many similar cases: 30,000 hectares of Procana land in Mozambique, 43,000 hectares of sugarcane land in Cambodia and so on. We argue that what is non-operational or what has failed in these cases is the promised capitalist enterprises; the act of land grabbing succeeded perfectly.

Second, the dominant method has difficulty accounting for and explaining the many cases of non-operational and failed land deals that were actually intentional: in other words, they were carried out exactly as planned, where land grabbing was the goal. As examples of this we can cite the strategies of 'timber mafia' in Southeast Asia and 'credit mafia' in Ethiopia. In Myanmar, only about one-fourth of all corporate land concessions saw capitalist enterprises emerging, such as oil palm, rubber and sugarcane plantations. In the southeastern part of the country, in the Tanintharyi region, many land concessions were granted supposedly for oil palm and rubber plantations. But in reality, the companies that were granted the concessions were never interested in developing agribusiness enterprises: they were interested only in extracting timber from the concession land. Having secured the concession contract, they proceeded to clear-cut the forest; many then planted a narrow row of oil palm trees, especially in concession areas bordering major roads, as a thin veil to hide their real motive of timber extraction.

In Ethiopia, the first wave of government land investment promotion that started in 2008 was two-pronged: targeting large-scale foreign and domestic corporate investors to acquire big tracts of lands, up to tens of thousands of hectares in size; and purportedly promoting medium and large-scale individual capitalist farms. In this second category, land concessions of between 100 and 1,000 hectares were granted to individuals who were then encouraged and supported to get loans from the EDB at quite favourable terms.

In Gambella region, several hundred acquired land and credit through this scheme. But once they had the long-term land lease contract, they proceeded to clear-cut the forest in the concession area, took the loans from the EDB, and disappeared. According to numerous interviews with regional and local officials and villagers, many of these concessionaires spent their loans on other things, in other places, such as buying apartments in Addis Ababa. It appears that this first generation of individual elites who got land concessions in Gambella were interested not in developing farmland as officially promised, but in securing money from the EDB. Now the EDB has foreclosed and taken over much of this land. In both cases of Myanmar and Ethiopia, land deals and land grabbing were not a failure; they were successful for those who received the land and the loans. The 'non-operationalization' of enterprises in these concession areas was not because of the failure of such capitalist enterprises, but because of the success of the original intention and the scheme—or rather, scam—of the beneficiaries.

Third, the implicit devaluing of the terms 'non-operational' and 'failed' land deals highlights a methodological flaw in the dominant literature on land grabbing, that is, a focus on plot by plot data-banking and analysis of land grabbing. This inadvertently leads to a blind spot when it comes to the system-wide logic that underpins land grabs, especially those associated with a convulsive land rush. It is akin to seeing the trees but not the forest. In our view, non-operational and failed land deals are intrinsic processes and outcomes of land grabbing and especially land rush.

The logic that produces operational land deals is the same as the one that produces non-operational and failed land deals. Non-operational and failed land deals play a role in conjuring a spectacle that is necessary to generate a supply of land and investments. Initially, land and investment prospectors do not know the optimal level for the supply of land and financial investment, leading to a tendency to play it safe by engaging in speculation and spectacularization. Once the supply of land and investment finance has stabilized, the inevitable outcome of these speculative and hyperbolic processes is an oversupply of land and, the flipside of this, an undersupply of investment finance. Therefore, non-operational and failed land deals on the one hand, and operational land deals on the other hand are co-constitutive: we can only understand one by understanding the other.

Recommended reading

Broegaard, R. B., Vongvisouk, T., & Mertz, O. (2022). The impact of unimplemented large-scale land development deals. *Frontiers in Sustainable Food Systems*, *6*, 789809.

Non-(re)distributive reform

The strategy of land redistribution through land occupation by the agrarian movement MST in Brazil is called land reform (Wolford 2010). Land redistribution carried out by the central state of China in 1949 and in 1978 is called land reform (Ye 2015). The buy-and-sell scheme promoted by the World Bank over the past four decades is called land reform (Deininger 1999). Yet, it is clear that land policies labelled as 'land reform' often have quite different political outcomes as to 'who gains and who loses' in the process. We can distinguish two types of land policies, namely, (re)distributive and non-(re)distributive reforms. In a separate entry we explain what qualifies a land policy to be labelled as (re)distributive land policy. In this entry, we describe non-(re)distributive reform and provide some examples.

The defining character of non-(re)distributive reforms is the maintenance of a status quo often characterized by land-based inequity and exclusion. The most typical and most common land policy in this category could be labelled the 'no land policy' policy which, in conditions of land-based inequities and exclusion, supports the existing distribution of land-based wealth and power. In other settings, a similar effect may be created when an existing land policy, even a redistributive land reform policy, is kept dormant 'from above' or becomes frozen or flounders in the course of implementation as it comes up against obstacles within the state or in society or both. However, this kind of situation should not be confused with others involving active land policies that are categorically non-(re)distributive, to which we turn next.

The most significant and widespread example of non-redistributive land policy is the 'formalization of land rights' framework, based on what we argue is a flawed assumption that poor people are poor because they do not have formal titles to their land (De Soto 2000; World Bank 2003)—hence the need to formalize land titles. Formalization of land rights without redistribution of wealth and power is essentially a formalization of inequality. Formalization of inequality occurs when an ahistorical, gender-blind, race/ethnic-blind and class-blind 'formalization' of land rights campaign is carried out in agrarian societies that are marked by socioeconomic inequality and lopsided power relations between various groups and classes. Formalizing land rights of legal claimants in settings with a high degree of inequality is likely to favour socioeconomically and politically dominant claimants. Depending on the society in question, these can be landed classes, agribusiness corporations, corrupt government officials, military and paramilitary groups, or indeed, the central state. In such cases, formalization policies serve only to formalize inequality and institutionalize historical injustice. Many private land titling programmes carried out by former colonial powers dispossessed the local population and facilitated land-grabbing by colonizers through formalization campaigns. In the contemporary period, formalization of land rights may have similar effects to the earlier waves of enclosures.

Market-assisted land reform (see separate entry, this book) is an emblematic example of a non-redistributive land reform. It represents a straightforward buying and selling of land which involves no significant change in the relative shares of the social groups involved; if there is any change, it is likely to be to the advantage of the elite seller of the land, as most cases of such land policies have shown in different countries (Borras 2003).

Land restitution is also non-redistributive in character if the entities that grabbed the land from villagers or the entities that have benefited from taking over the land of villagers who were displaced from their territory for various reasons, do not lose their control of the land, and if those who lost the land never regain access and control. In practice, most land restitutions fall within this category. The notion of land restitution is based on the idea that a mass of people and not just some random individuals, either lost their land through direct stealing by a dominant social group or were forced to abandon their land, usually because of violent conflicts. What these people have lost is their means of production and social reproduction—land centrally. In this context, land is a productive resource (farmland, grazing land, etc.), a landscape and agroecological zone, territory and homeland. Land in social reproduction is key to labour and generational reproduction and to socio-ecological, cultural and political reproduction. What is also lost is the community.

There are currently tens of millions of internally displaced people in the world; one of their main demands is the restitution of their lands. But land restitution as a land policy is among the rarest of all land policies. In the contemporary era, in the few countries which have a policy of land restitution, the meaning most commonly accorded to the term restitution is 'compensation'—compensation for the earlier displacement and expulsion of people from their land. This is then monetized and paid out by the state, as in many cases in South Africa. In Colombia, land restitution has been conducted through 'individualized land claims and awards'. This is perhaps one of the most absurd interpretations of what land restitution is and ought to be. It is true that an individual's home lot and farmland are central, but disembedded from the broader reality of which these were a part—that is, land in production and social reproduction sustained by the integrity of a landscape and agroecological zone—the restitution of land individually to selected legally qualified individual claimants can be meaningless. Without restoring the destroyed community landscape and the social and physical infrastructure that were there before, the policy is effectively sending individuals into hostile conditions where the likely outcome is that land restitution beneficiaries either do not claim the land or simply sell it. Moreover, this approach to restitution does not entail a loss for those who previously benefited from the land grab or caused the displacement of the people (see, for example, Otsuki 2023). The context and examples discussed here demonstrate the conditions under which a land restitution policy is in essence non-(re)distributive.

Elsewhere in this book, we discuss the popularity of share tenancy reforms in the twentieth century and, to some extent, in the contemporary period and outline the conditions under which these can be truly redistributive in character (see '(Re)distributive land reform', 'Land tenure security', this book). However, share tenancy reform can also be a straightforward formalization of existing tenancy arrangements without any redistributive content, merely ratifying what exists. This is most commonly seen in share tenancy reforms that are limited to a shift from informal to formal contracts, where the only substantive change is, for example, to ensure that the landlord cannot unilaterally abrogate the contract without due process. Such measures essentially formalize inequality and lock the tenant-farmer into the exploitative terms of a tenancy contract.

Finally, there is a trajectory that can be termed 'counter-reform' that also qualifies to be included in the non-(re)distributive reform category. The convention of resettling potential and actual land claimants to 'empty public lands' may, under certain conditions, have some potential for redistribution, although historically it has impacted negatively on affected pre-existing settlements of local populations (Scott 1998: 69). However, where such a resettlement policy is carried out precisely to avoid and undermine political agitation for redistributive reforms in the larger agrarian society, then in effect it constitutes a counter-reform. The series of so-called 'transmigration' land policies in Indonesia is a good example.

Our main point is that different people use the same terms—land reform, land restitution, and so on—but mean completely different things. Framing land policies in the context of redistributive and non-redistributive policies is a useful way to guide our analysis as to 'who gains, who loses', and avoid the mistake of treating very different land reforms in diverse societies and moments as if they are all the same.

O

Operational land deals

The concept of 'operational land deals' (or 'operational land acquisition') is associated with the massive literature on land grabbing that has accrued since 2008. It has to be read alongside its opposite, the concept of 'non-operational land deals' (see separate entry, this book). At first glance, it seems like a straightforward term needing no more than a sentence or two of explanation. But in the context of current scholarship and public policy debates, it actually requires further clarification.

The concept began to be used more frequently in the 2010s. A few years after the first widespread media reports about the global land rush, it became clear that many of the large land investments which had been announced through press releases by governments and companies had failed to progress into full-scale, concluded land investments. Moreover, many of the promised capitalist enterprises that had been planned for the lands acquired by investors through land deals were not pursued or were initially pursued but later stalled or were cancelled for various reasons. And so the terms operational and non-operational land deals gained currency.

One trend in the literature on land grabs, especially in determining the character and the scope of land grabs, is to weed out other data and focus mainly on what is considered a clean dataset, that is, information related to operational land deals. In the most basic definition, operational land deals are deals that resulted in something concrete (see, for example, Cotula et al. 2014; Hall et al. 2017). For the framers of the Land Matrix, the most consequential database on contemporary land grabbing (Anseeuw et al. 2013; Nolte et al. 2016; Lay et al. 2021), large-scale land acquisitions are defined as those land transactions that have foreign and domestic investors, involve at least 200 hectares, and were contracted not earlier than 2000. The Land Matrix works with four classifications of land acquisition status, namely: (1) 'Concluded', for contracts that have been formalized, with subcategories (a) not yet started, (b) startup phase, (c) in operation, (d) abandoned, and (e) 'none', meaning that no information is available; (2) 'Failed', which means the deal was started but abandoned; (3) 'Intended', for deals that have neither

DOI: 10.4324/9781032658438-11

concluded nor failed; and (4) 'Other', which includes two subcategories: (a) contract expired and (b) change of ownership (Borras et al. 2022).

Subsequent analyses and use of the dataset do not always follow the Land Matrix categories exactly. Instead, terms and concepts get simplified, fall in a grey area or become muddled. For example, a 'failed' land deal is loosely interpreted to refer to one of several possibilities, such as failure to strike a formal land investment agreement; a successful formal land concession but failure to actually take possession of the land; successful acquisition of the land but failure to build a capitalist enterprise; and so on. The same is true of 'operational land deals'. The simplified, more common understanding of this term encompasses those formally concluded land investment contracts with various stages of capitalist enterprise construction: not yet started, in the startup phase and in operation.

The concept of an operational land deal is a useful one, however imperfect. It is at least a measurement of something. However, it could be clarified, improving its potential usefulness even if inherent limitations remain. In this context, it is important to point out a number of conceptual problems with the current dominant view of what the term 'operational land deals' means. First, within the parameters of the Land Matrix, it is possible that at least some of the land deals included in category 4 ('Other') are in fact operational land deals; this seems likely to apply especially to the subcategory 'change of ownership', and arguably, may also apply to at least some in the sub-category of 'contract expired'. These two subcategories are formalistic status categories which may not affect the nature of the land deal and the capitalist enterprise.

Second, there is one particular type of land grab that does not fit within the defining features established by the Land Matrix and which cannot be captured by the crowd-sourcing method used by database initiatives like the Land Matrix or even by particular official census data by governments. This is the 'pin prick land grabs' category (see separate entry, this book). An individual land grab can be small in scale in relative terms which would never attract media attention and would not be considered worthy of reporting. Many of these small-scale grabs are done informally and by stealth and are unlikely to qualify for official census-tracking methods. But these scattered, small-scale instances of land grabs that at times are triggered by a land rush can be quite widespread in a particular region, and when aggregated, their scope can be at least as extensive as those large-scale corporate land deals that are recorded. This is discussed in greater depth in the entry 'Pin prick land grabs', this book.

Finally, if we define land grabbing as 'control grabbing', we should consider institutional mechanisms such as contract farming as operational land deals—and indeed the latest iteration of the Land Matrix includes contract farming. If we follow this logic, then we also need to include trade-based land control grabs, especially when the buyer has established a monopoly or near-monopoly in a global commodity chain. For example, while the majority of soya is grown in the Americas, buyers from China control up to 70% of the international soya trade. Xu and Borras (2024) calculate that China's purchasing power in the global soya commodity chain can be translated into an effective control of distant land use equivalent to more than 30 million hectares of land. If our interest in tracking operational land deals is to see how land use change comes about and how benefits are extracted from such land use dynamics, then social change inherent in indirect land use change (ILUC) is not very different from the many examples of operational land deals discussed in the literature. Combining views on direct land use change with ILUC, especially in the context of the geopolitics of land becomes an imperative (see, for example, Dwyer 2022).

Ultimately, the usefulness of the concept of operational land deals—whether engaging with parts of it, as is currently the norm in the literature, or with the idea of an enlarged whole, as suggested in the discussion here—depends on the specific research question one is addressing, or the political debates one is involved in. In both cases, it is important to understand the concept of 'operational land deals' in relation to 'non-operational land deals' (or 'failed land deals); see separate entry, this book.

P

Petty reformism

We define 'petty reforms' as reforms that are small in scale but are concrete enough to make a relatively positive impact on the lives and livelihoods of working people who need such positive changes in their everyday life. Scale here is relative. A reform that impacts several households in a village is small in scale for analysts who are interested in seeing positive changes beyond the level of a village but may not be seen as 'small' by the households in the village that benefit.

In the context of the politics of land, by petty reforms we mean concrete but small-scale positive changes in everyday land-based social relations. This might be a change in a sharecropping arrangement, with some increase in the share of the tenants or sharecropper, or it might be a number of villagers regaining access to their community forest after the lifting of some prohibition imposed by the state. It could also mean tens of thousands of households receiving land through a redistributive land policy, in the context of millions of households needing land. Petty reforms in land politics do matter to working people, who may not see them as small. Therefore, it is important not to be dismissive of such small-but-concrete reforms that, separately and together, are significant to working people.

Nevertheless, if our concern is for wider, more transformative positive changes in the lives and livelihoods of working people, then we should not limit our intellectual curiosity and political interest to small-scale changes in local communities. Rather, it is important to see how wider reforms, even system change, can be accomplished. This is not an either/or question, that is, either small reforms now, or system change later; rather it is both. Rejecting system-wide positive changes on the grounds that these are not feasible in the current conjuncture is defeatist.

As Eric Olin Wright (2016: 102, original emphasis) reminds us:

> Pessimism is intellectually easy, perhaps even intellectually lazy. It often reflects a simple extrapolation of past experience into the future. Our theories of the future, however, are far too weak to really make

DOI: 10.4324/9781032658438-12

> confident claims that we know what can't happen.... The appropriate orientation towards strategies of social transformation, therefore, is to do things now which put us in the best position to do more later by working to create those institutions and structures that increase, rather than decrease, the prospects of taking advantage of whatever historical opportunities emerge.

While we support positive, concrete local reforms in land politics now, we are also conscious of the danger of these reforms being 'merely agrarian', that is, too localized, scattered and sectoral. This builds on Nancy Fraser's discussion of climate politics and, specifically, on her critique of existing environmental struggles as being 'merely environmental', implying that they are unable to connect to broader class struggles within a society in order to become anti-systemic, anti-capitalist struggles (Fraser 2021). Fraser argues that 'to become counter-hegemonic... a new commonsense must transcend the "merely environmental"', and furthermore:

> it must connect its ecological diagnosis to other vital concerns—including livelihood insecurity and denial of labour rights; public disinvestment from social reproduction and chronic undervaluation of care work; ethno-racial-imperial oppression and gender and sex domination; dispossession, expulsion and exclusion of migrants; militarization, political authoritarianism and police brutality.... Only by addressing all major facets of this crisis, 'environmental' and 'non-environmental', and by disclosing the connections among them, can we begin to build a counter-hegemonic bloc that backs a common project and possesses the political heft to pursue it effectively.
>
> (Fraser 2021: 96)

We see the tendency for our intellectual and political interest to get stuck at the level of small reforms as detrimental. This brings us to the issue of 'petty reform incrementalism' or 'petty reformism'. We define 'petty reformism' as the political consciousness, or ideology, that justifies limiting the parameters of political struggles and reforms to what is do-able within a given correlation of state and social forces, with a specific social issue as the sole objective. In other words, small reforms *in lieu of* system-wide reforms. Thus, instead of small reforms serving as building blocks towards wider reforms in the future, they become the end goal.

Globally, we fear that the politics of redistributive land policies is currently stuck in petty reformism. Concretely, land reconcentration in the hands of the few has gained momentum, and in many places, land concentration and land monopoly either have not been dismantled (such as in Colombia and South Africa), have been increasing (such as in Myanmar and Cambodia) or have recently re-emerged (such as in Europe). In such societies, what is needed are transformative, system-wide redistributive land policies; by contrast, actual state policies and programmes, as well as the advocacy positions of civil society organizations, are far too small-scale and fragmented. This can be illustrated by a few hypothetical scenarios.

First, in a context in which the actual demand for land is by and for 5 million landless working people, the government land programme undertakes to buy and sell land from selected beneficiaries of 100,000 households. Second, the actual demand for land restitution is by and for a displaced population of 1 million; the government initiates a programme to resettle 10,000 displaced peoples. Third, the actual demand for land redistribution is by and for 3 million working people; the government programme merely formalizes the existing land occupancy of 20,000 households. We could go on enumerating more hypothetical cases that are in fact far too common and too real in the contemporary world, but three are sufficient for our purposes. The reforms in these three illustrative examples are not treated as initial building blocks that could inspire and open wider reforms in the future; rather, they are used as justifications to slow down and eventually stop and block further reforms. These are small reforms now in lieu of wider reforms both now and in the future. This is a concrete illustration of what we call petty reformism.

This political mindset is partly an outcome of neoliberalism and the waning influence of communist and socialist political parties. Organization-focused work has taken over from class-wide contention as the more common type of activity, its advance coinciding with the global rise of NGOs. For example, land struggles are more likely to be launched on behalf of members of specific organizations, while demands for class-wide structural reforms have dwindled. There is a material basis for this. NGOs receive funds from donors; donors in turn get their funds from western governments or religious fundraising campaigns, or both. They are expected to account for how the funds are spent. Hence, the projects that are most likely to be funded are those that can deliver concrete outcomes in terms of improving the lives of working people, usually in the Global South. These outcomes have to be measurable and timebound, showing, for example, how many households actually received lands within the three-year project period. NGOs and farmers' movements therefore mobilize to secure land for their members or would be members. Campaigns for system change, such as the redistribution of land to all landless and near-landless, become too vague, too difficult to measure and not feasible within a project's short duration. Thus, we see the proliferation of these kinds of limited land campaigns, rather than calls for class-wide and system-wide redistribution and change.

Organizational work and political mobilization have become too parochial, 'too localist'. This is not to say that localized mobilizations and claim-making campaigns focused on the members of particular organizations are not important; they are. In fact, one of the weaknesses of conventional left agrarian movements of the national liberation generation was that they tended to be too concerned with strategic and long-term goals at the expense of the concrete and urgent needs of rural working classes. This was often coupled with an 'all or nothing' perspective (that is, complete system change or nothing), while tactical fights around incremental reforms were shunned. Now, however, the pendulum seems to have swung too far in the other direction: in the era of neoliberalism, localized gains for organization members are favoured above system-wide struggles for the benefit of the working classes more generally. In this context of striving for what is 'do-able', what can be gained without recasting the existing balance of social class forces, it has become less common to see struggles that combine tactical with strategic goals, organizational membership demands with class-wide struggles.

The degree of land concentration and land monopoly is often too great for localist and NGO project-based campaigns to make any significant impact. System-wide transformative reforms are urgent and necessary—reforms that have a cumulative, ratchet effect towards greater structural reforms or even revolutionary social change in the future. These reforms have to be subversive and insurgent enough to unsettle the status quo; they are the opposite of petty reformism, which is anti-insurgent and anti-subversive.

Recommended reading

Wright, E. O. (2016). Socialism and real utopias. In R. Hahnel & E. O. Wright (Eds.), *Alternatives to capitalism: Proposals for a democratic economy* (pp. 25–40). New York: Verso.

Pin prick land grabs

We argue that corporate land deals, both operational and non-operational, in the context of a land rush are often accompanied by—or trigger—a parallel, third category of land transactions, namely, small-scale instances of land accumulation that are usually non-corporate and often outside the media and academic research spotlight. We call this category 'pin prick' land accumulation (Borras and Franco 2024a). It usually entails small-scale, scattered, often stealthy and almost invisible instances of land accumulation, which, when aggregated, can become widespread, ubiquitous and large-scale, both in terms of capital involved and in geographic terms (Borras and Franco 2024a). The notion of scale here is relative: a small scale could mean 2 hectares or 1,000 hectares of land depending on local social structures and institutions and the conditions of an unfolding land rush. The notion of stealth here is also relative: pin prick land grabs are all too obvious for those in the local communities where they are happening, but thanks to their scattered nature, these individual instances may not be apparent to outsiders. And yet, put together, these small-scale instances of land accumulation may have the same effect as much larger-scale corporate land grabs: they often change land use and land control extensively.

Pin prick land accumulation often happens without the direct involvement of corporations, most commonly driven by powerful individuals or non-corporate groups. In many communities which have experienced this category of land grabbing, the resulting agrarian and ecological transformations can be far-reaching. They might include, for example, transforming customary swidden agriculture-based agroecological zones into monoculture landscapes based on individual property and farm operation or transforming a vast grazing area of meadows and forest of a nomadic pastoralist community into a single monoculture or a collage of small and medium capitalist agricultural or industrial forest enterprises. Where these pin prick land accumulations and grabs happen, they can undermine the integrity of the affected landscape or agroecological zone, the basis of human and non-human life. Pin prick accumulation is essentially the death of agrarian communities by a thousand cuts (Borras et al. 2024).

Pin prick land accumulation happens all the time and everywhere under the generalized commodification of nature and land within global capitalism. This type of everyday accumulation has been well studied in agrarian studies, but here, we focus specifically on the pin prick land accumulation associated with the contemporary land rush, which is in turn associated with financialization of agriculture, food and nature (Ouma 2016; Clapp and Isakson 2018; Ouma et al. 2018; Fairbairn 2020). This category is under-studied in the contemporary scholarship on land grabbing, with only a few exceptions, including Beban and Gorman (2017), Borras et al. (2020a), Friis and Nielsen (2016), Hilhorst et al. (2011), Kandel (2015), Woods (2020) and Xu (2018) (see Borras et al. 2022).

Stepping back and looking at the broader perspective and longer trajectory of land politics, our argument is that a land rush necessarily generates three streams of land accumulation or land grabs, namely, corporate land deals that are operational, corporate land deals that are non-operational (popularly referred to as 'failed' land deals) (see separate entries, this book) and pin prick land accumulation or grabs. The first category attracts the global spotlight, with the second also increasingly subject to scrutiny. Meanwhile, the third category often happens outside the spotlight of media and academic attention, partly because these instances of land grabbing—or 'control grabbing'—are not as easy to observe or as dramatic; they do not involve well-known corporations or corporate personalities and are not often brokered by the central state. Pin prick land grabs are generally not legible enough for large-scale data banking initiatives to be able to identify and quantify, and may not be interesting enough to others because they do not involve well-known faces of global capitalism.

We further argue that the three categories of land transactions (operational, non-operational and pin prick land grabs) not only exist simultaneously, but that they are *co-constitutive* of one another in the context of a land rush. We differentiate 'land grabbing' from 'land rush' (see separate entries, this book). For us, land rush is that 'chaotic, relatively short-lived historical juncture marked by a sudden surge in demand for land, accompanied by an extremely speculative and competitive, often violent and convulsive transition from one set of rules on commodity and land politics to another' (Borras and Franco 2024a: 3). At the start of a land rush, neither land prospectors nor investment prospectors (see 'Land prospecting', this book) are likely to know exactly what is needed in terms of quantity of land, level of capitalist investment, where and when. It is partly this uncertainty that triggers the combined speculation and spectacularization that cause key actors in the land rush to try to get ahead of everyone else to acquire lands or to secure investors. There is thus a tendency to deliberately make hyperbolic projections about land supply and investment possibilities that ultimately lead to many of the projected land deals fizzling out once the reasonable and feasible level and extent of land and investments start to settle. This explains why many projected land deals are (inevitably) not pursued, and why, for those land deals that are initially concluded, many planned enterprises are bound to be cancelled, withdrawn or scaled down.

Another effect of land rush is that a wide array of actors, often non-corporate, join the bandwagon, the fevered frenzy, of the land scramble. They are individual scammers, swindlers, speculators, brokers and entrepreneurs who are attracted to and seduced by the same promise of a windfall that attracted corporate participants to the land rush. They can act legally or illegally, openly or by stealth, with consent or by force. These are the pin prick land grabs. Pin prick land transactions include distress sales, land brokering, theft, coercion, swindling, contract grower arrangements, etc., and are a key element of the frenzied land rush, yet they have rarely been examined in the context of the latter.

Pin prick land transactions in the context of a land rush can be understood in several ways. First, in capitalism, the commodification of nature, especially land, is underway worldwide, albeit spatially and temporally unevenly. Before the surge that defines a land rush, there are already pin prick land transactions. These are then scaled up in the processes of a land rush. Second, state-enabled and directed commodification of land can occur through large-scale corporate or small-scale non-corporate processes; both of these categories achieve a faster pace and a wider reach during a land rush. Third, a state-initiated commodification of land through large-scale land grabs is watched by, and triggers speculation, not only among the targeted corporate actors but among the general public as well. Particular individuals are lured by the spectacle of the land rush (Tsing 2000), and join the process in various capacities, as land buyers, speculators, brokers, moneylenders, adjudicators, surveyors, map makers, cadastre recorders, con artists who make fake documents and notaries, and all sorts of scammers, swindlers and thieves. Some of these aspects have been documented and examined by Sud (2014) on land brokers and Levien (2021) on land mafia in India, and by Kröger and Margutti (2024) in Brazil. The resulting deals fall into the category of pin prick land transactions.

Recommended reading

Borras Jr, S. M., Arango, L., Belay, M., Franco, J. C., Kham, S. S., Moreda, T., Rojas, I., Wang, C., Ye, J., & Xu, Y. (2024). Death of agrarian societies by a thousand cuts: 'Pin prick' land grabs and the land rush. LDPI Working Paper Series No. 43. https://www.peasantjournal.org/. Downloaded 14 August 2024.

Political economy

We clarified in the beginning of this book that our take on the politics of land is from the perspective of Critical Agrarian Studies (see 'Critical Agrarian Studies', this book) which is, in turn, fundamentally based on radical political economy. Our understanding of radical political economy speaks to how Henry Bernstein has defined it, which is broad enough to accommodate different currents within the intellectual tradition. For Bernstein, political economy is the field of study about

> the social relations and dynamics of production and reproduction, property and power in agrarian formations and their processes of change, both historical and contemporary." Understanding agrarian change in the modern world centres on the analysis of capitalism and its development. By capitalism I mean a system of production and reproduction based in a fundamental social relation between capital and labour: capital exploits labour in its pursuit of profit and accumulation, while labour has to work for capital to obtain its means of subsistence. Beyond this initial and general definition, and indeed within it, there are many complexities and challenges.
>
> (Bernstein 2010: 1)

For Bernstein (2010: 22–24), there are four fundamental questions in political economy. First is 'who owns what', which is the social relations of the means of production and social reproduction, specifically concerning about its distribution across property regimes, namely, private, state, commons and hybrids in between. This means it is important to understand the politics of distribution of access and control in the context of a 'mosaic' of property regimes, and paying attention to the changing configuration of the mosaic under conditions of generalized commodification where state and common lands are increasingly commodified and privatized. Second, is 'who does what' which is the social relations of property. Depending on the character of social relations of property, individuals associated in a group, for instance, those with small plots of land in the countryside, are likely to engage in broadly similar productive and social reproductive activities in order to survive or even accumulate. In this process, they engage with other social groups, including those who do not produce, such as the landlord class. Third, is 'who gets what' which is the social division of income, remuneration, rent or profit, whether by the smallholder farmer or the landlord. Fourth, is 'what do they do with it' which is the social relations of consumption, reproduction and accumulation. Bernstein explained that "this final question is about how different social relations of production and reproduction determine the distribution and uses of the social product" (Bernstein 2010: 23). Bernstein concluded:

> These four key questions can be usefully applied across different sites and scales of economic activity, from households to "communities" to regional, national and global economic formations. They can also be applied to different types of societies at different historical moments. There is also an implicit sequence in the four questions: social relations of property shape social divisions of labour, which shape social distributions of income, which in turn shape the uses of the social product for consumption and reproduction—which, in the case of capitalism, includes accumulation.
>
> (Bernstein 2010: 24)

A key point we want to make in this book is that it is not possible to fully grasp the politics of land, usually seen from Bernstein's first question, without understanding how it is embedded in the other three fundamental questions about social relations of labour, income and remuneration, as well as consumption, reproduction and accumulation, or ultimately, of dynamics of capitalism. Other scholars have suggested the importance of looking at a fifth question, that is, how do these four questions shape and are shaped by the question of political ecology. A bottom-line in political economy as briefly defined here is the centrality of class analysis (see 'Class', this book).

Recommended reading

Bernstein, H. (2010). *Class dynamics of agrarian change*. Halifax, NS: Fernwood Publishing; Rugby: Practical Action. For Open Access ebook edition: https://practicalactionpublishing.com/book/2571/class-dynamics-of-agrarian-change

Property

> A (usually material) thing belonging to a person, group of persons, etc.; a possession; (as a mass noun) that which one owns
>
> Originally: a piece of land under one ownership; a landed estate. Now also: any residential or other building (with or without associated land) or separately owned part of such a building (as an apartment, etc.).
>
> (Oxford English Dictionary n.d.)

The most common understanding of 'property' is as a 'thing', as the Oxford English Dictionary definition shows. Our treatment of property is less about the 'thing', and more about the social relations based on it. As Anna Tsing (2002: 97) reminds us, 'property is a social relationship between nonowners and owners, in which nonowners are expected to respect the rights of owners to their claimed objects'. It is 'an enforceable claim of a person to some use or benefit of something' (Macpherson 1978: 4). There are four fundamental questions in critical political economy, as explained by Henry Bernstein (2010), namely, who own what? (social relations of property), who does what? (social relations of labour), who gets what? (social relations of income and remuneration) and what do they do with the wealth or surplus generated? (social relations of consumption and reproduction). These questions are all about social relations, and the social relations around property are perhaps the most fundamental, shaping what one can or cannot do for a living. In agrarian political economy, land property is one of the most important types of ownership, often referring to agricultural farmland and its relationship to the development of capitalism (or construction of socialism) more generally.

Property has not always been understood as a 'thing'. Before the full development of global capitalism, a more common understanding of property was the right in something. Macpherson (1978: 7) explains that the change in the common meaning of property 'came with the spread of the full capitalist economy from the seventeenth century on, and the replacement of the old limited rights in land... by virtually unlimited rights'. Where this happened, and 'as rights in land became more absolute, and parcels of land became more freely marketable commodities, it became natural to think of the land itself as the property' (1978: 7). From this point onwards, 'the distinction between the right and the thing has become increasingly blurred' (1978: 7).

Property implies an enforceable claim, which means property is inherently political as it requires an actor who should enforce, and is relational as it implies a claim over a particular property in relation to others (Macpherson 1978: 4). Moreover, and following on the discussion above, when we say property, the 'thing' that we are talking about is often understood as 'private property'. Property making is equated to private property making. Narratives about property tend to stress the merits of private property, as private individual land titles have become hegemonically understood to be intrinsically linked with security, as in secure land rights, land tenure security. This is problematic.

For Macpherson (1978), there are at least three broad types of property, namely, private, state and common. Private property means that someone has rights to exclude others from something (1978: 4). This can manifest in various ways, depending on the institutions that created and maintain such rules on enforceable claims. It can mean freehold, alienable private property rights over a plot of land, or individual usufruct rights over such a plot. Ostrom (2001: 129) criticized conventional theory on 'idealized models of private property and government property'. She argued that, from such a traditional perspective, 'the concept of private property is conceptualized narrowly with a primary focus on the right to alienate through sale or inheritance as the defining attribute of private property' (2001: 129). Ostrom explored the different types of claims made over different types of rights over various resources.

These rights include the right of access, withdrawal, management, exclusion and alienation. Only a full owner has the complete set of enumerated rights, while other types of claimants have varying rights or combinations thereof, but not the right of alienation (2001: 135).

According to Macpherson (1978: 4), common property is 'the right of each individual not to be excluded from something'. He elaborates further:

> Society or the state may declare that some things—for example, common lands, public parks, city streets, highways—are for common use. The right to use them is then a property of individuals, in that each member of the society has an enforceable claim to use them. It need not be an unlimited claim... but the right to use the common things, however limited, is a right of individuals.
>
> (1978: 4)

Thus, both private and common properties imply the right of individuals, despite the term 'common' at first glance suggesting otherwise. These individuals can be both natural persons or artificial persons, for example, corporations.

The third type of property is state property. State property is not a common property, but rather an 'exclusive right of an artificial person'; in this case, the artificial person is the state (1978: 6). State property 'consists of the rights which the state has not only created but has kept for itself or has taken over from private individuals or corporations'.

In between these ideal types of property, there are various property types that combine elements from the three broad types, such as usufruct rights, customary land tenure, collectives, and so on. However, any attempt to capture the diversity of social relations around land through typologies, as Macpherson does, is bound to generate debated and contested outliers. For example, an updated version of the discussion by Macpherson on common property would need more nuance, especially from the perspective of vibrant research about the commons and common pool resources (see 'Commons', this book). Nevertheless, the two points that we wish to emphasize here are that property does not mean only private property, and that it is not a thing, but rather a set of social relations.

Recommended reading

Macpherson, C. B. (Ed.). (1978). *Property, mainstream and critical positions*. Toronto: University of Toronto Press.

Public lands

The term 'public land' is quite vague, and is often used alongside terms such as 'state land' and 'government-owned land'. To add to the unclarity, what observers are referring to may differ from one society to another, from one period to another. We might try to define the term by what it is not, namely, land that is not legally classified as private property. However, this only takes us so far, as there are different legal categories of lands that are not privately owned (see relevant discussions in Alden Wily 2012; Kelly and Peluso 2015).

Many 'open access' lands and 'commons' (see 'Commons', this book) fall under the heading of public lands. Some non-private lands were reclassified as 'government-owned lands' with various official purposes attached to them (e.g. educational, penal, resettlement). Most of the lands declared as 'reserves', such as forest reserves, wetland reserves, national parks and other conservation sites, are also in the public land category. Different branches of government decide on the fate of these different categories of so-called public lands, whether they can be alienated, disposed of, reallocated for private ownership and use, and so on. Some types of public land require the legislative branch to pass a law before they can be reclassified, others require a court intervention, while most others fall under the authority of different ministries within the executive branch of government.

Here, we will briefly discuss a number of issues that are relevant in the context of Critical Agrarian Studies, the field in which we locate this book.

First, one implication of lands being classified as non-privately titled lands, and thus, 'public lands', is an almost automatic assumption that they are owned by the central state—an assumption that is used by the state to justify its claim over these categories of lands. An extreme interpretation of the notion of non-privately titled land—or what John Locke called 'non-enclosed' lands—is to assume or suggest that such land is empty, vacant, a wasteland. During the global land rush of the twenty-first century (see 'Land rush', this book), one justification for the idea of large-scale land reallocation for big-time investments has been this claim by governments and international institutions that there are large areas of marginal, empty, unutilized and available land. The 'availability' element of this argument stems from the fact that these lands are not privately enclosed (meaning, not privately titled); they are therefore claimed by the state as public lands, giving the state—and the state alone—the power to decide whether such land can be reallocated to other land users for other land uses, and if so, how. The assumption that these lands are empty is taken almost straight from Locke's text (Locke 1978).

Second, however, many of the lands claimed and classified by the state as public lands are not actually empty and unused. These are lands utilized by social groups—usually Indigenous peoples—and especially those whose methods of usage do not necessary involve sedentary settlement and production. For example, a plot of *taungya* land in Myanmar may be used for sedentary production, but the cultivator lives in a distant village, while others who use *taungya* land employ mobile cultivation techniques. This entails leaving some land fallow. The existence of fallow lands then becomes a justification for claiming that the lands are unused, unenclosed, and can therefore be categorized as a 'wasteland'. The Vacant, Fallow and Virgin Land Management

Law of 2012 in Myanmar, or VFV Law, does exactly that. The government used this law to ask everyone to register their claimed lands; if they are deemed eligible, claimants get a plot of land; in cases of mobile cultivation, this is often just a portion of the whole range of land they use. Lands that are unregistered are then categorized as 'vacant, fallow and virgin' lands, claimed by government and reallocated to big owners of capital for large-scale agribusiness production or mining operations (TNI 2018). It can thus be seen as a government strategy of using formal land titling in order to release large areas of land that is not privately enclosed and which the government then reallocates to medium- and large-scale land investors, both domestic and foreign.

Third, this method of classification of land is usually plot-oriented and production-centred. It requires claimants to produce documents that show their plot of farmland. This process invalidates many types of land access that are not plot-based and not necessarily for production, but for social reproduction. For example, a household is granted its request to get a plot formalized and titled, say, 5 hectares of land for farm production. But this does not include the household's layers of access to various types of land and nature, and for various productive and social reproductive activities that aim to produce both exchange value and use value (see, 'Land in production/social reproduction', this book).

Fourth, in the vast literature on land reform, public lands are usually overlooked and almost always excluded from accounts of land redistribution. The classic definition of land reform is the redistribution of big private estates to landless and near-landless peasants and farmworkers (Griffin et al. 2002 see also, 'Land reform', this book). The underlying assumption here is that land reform is a zero-sum process, and in the case of public land, no private entity is losing. This assumption is based on an uncritical acceptance of the definition of public lands as 'non-enclosed', and thus empty. This is of course a deeply flawed assumption, and easily debunked by countless empirical cases showing that in most societies, private control and interest in public lands are quite entrenched.

In the Philippines, some of the biggest landlords quietly forged various long-term lease arrangements with government over these public lands, at nominal cost. For example, public land categorized as 'timberland' is supposed to be demarcated and preserved for continuous forestry and timber conservation/production. In reality, many of these lands have been leased, usually for 25-year renewable terms, and are used for cattle ranching and agribusiness monoculture. Similar situations apply to other land categories like foreshore land and forest land. Other cases have been subject to complicated manoeuvres, such as the 30,000 hectare Davao Penal Colony (DAPECOL) in Davao, a vast portion of which is under long-term lease agreement with a powerful landlord-politician. One institutional instrument of land redistribution in the Philippines is Community-Based Forest Management (CBFM), which entails a 25-year contract with a community group. Although this is rarely considered part of any land reform process, we would disagree. In many of these cases, there has been real redistribution of political control over these lands (Franco and Borras 2005).

With the state having hegemonic claim over these lands, as well as monopoly of the coercive apparatus of the state, public land can and often has been used, across history and in many societies, to support the capitalist state's two permanent but contradictory tasks, namely, facilitating capital accumulation while trying to maintain a minimum level of political legitimacy (Fox 1993; O'Connor 2017 [orig. 1973]). In Colombia, for example, the state uses public land to facilitate medium- and large-scale capitalist accumulation by allocating land for individual *fincas*, and also for large-scale land acquisition. This is particularly clear in Altillanura, the epicentre of the land rush in Colombia (Coronado Delgado 2022; Arango 2023; Hurtado-Hurtado et al. 2023), but examples can also be found in many other countries, including Ethiopia, Mozambique, Cambodia, Myanmar and the Philippines.

We anticipate that public lands will remain a key battleground for land control internationally, especially because conservation groups are also targeting these lands for large-scale conservation projects, such as the 30 × 30 initiative (30% of the planet's surface land should be placed under conservation by 2030) (Büscher 2024) and various carbon offsetting schemes in relation to climate change mitigation and adaptation.

Recommended reading

Hurtado-Hurtado, C., Ortiz-Miranda, D., & Arnalte-Alegre, E. (2023). The legal contention for baldíos land in the Colombian Altillanura. *Latin American Politics and Society*, *65*(2), 55–82.

Oxford English Dictionary. (n.d.) https://www.oed.com/dictionary/property_n?tl=true.

R

Reconcentration

Landlessness, near-landlessness, and land monopoly are not only caused by land grabbing carried out through various forms of force, intimidation and violence, or through the dull compulsion of economic relations. These two drivers, separately and combined, have caused large-scale and widespread reconcentration of land. For example, the level of contemporary concentration of land control in the hands of a few in the European Union is extremely high; this has mainly occurred through market relations that are partly shaped by state policies on market prices and subsidies (van der Ploeg et al. 2015). However, concentration can also be an outcome of various types of land policies, including those that are generically labelled as 'pro-poor land policies'. Akin to the discussion on redistributive and non-redistributive reforms (see separate entries, this book), the concept of 'reconcentration' is useful in helping us understand 'who gains, who loses', especially when these outcomes are not obvious in land policies that are billed as 'pro-poor'. Here, the concept of 'reconcentration' and the mechanisms that enable it become important.

Land policies, despite pro-poor claims, can actually result in anti-poor outcomes by reconcentrating land control in the hands of the few. The defining character of land policies that result in reconcentration is that while land-based wealth and power transfers do take place, access to and control over the land resource actually gets (re)concentrated in the hands of the non-poor: private landed classes, owners of big capital, state or other elite community groups. This kind of change can occur in private or public lands. The organization of control over land resources can be through individual, corporate, state or community group institutional arrangements in property rights. The transfer may involve full or partial land ownership. Different variations are possible, but the bottom line is the same: those who benefit from land-based wealth and power transfers are landed classes and other non-poor entities or the state and those who lose are ordinary working people. There are at least three broad trajectories within the (re)concentration category.

DOI: 10.4324/9781032658438-13

First, *reverse redistribution*, is where previously redistributed land-based wealth and power (from the landed classes or the state to the working people) are later redistributed again, back to the landed classes, other elites or the state. This kind of reversal was seen in Chile after the 1973 coup by Pinochet, who returned a significant portion of land redistributed by the Allende government to its previous owners and other elite entities (Bellisario 2007). In addition to such large-scale reversals, there are also what can be called 'pin prick' micro reversals involving small units of landholdings that had previously been redistributed to peasants through land reform and other (re)distributive land policies. Individually small in scale, when taken together they constitute a significant trend in reversal of previous redistributive gains. This type of reversal can be seen today in settings that have experienced significant land redistribution or land tenure reform in the past, such as in West Bengal and in the Philippines.

Second, *perverse redistribution* is a trajectory in which land-based wealth and power are transferred from working poor people to the landed classes, other elites or the state or elite community groups. This can happen under a variety of generic land policies, including land reform, forest land allocation or management devolution, formalization and privatization of land rights, land-based joint venture agreements, land-lease arrangements, and so on (take case of land reform beneficiaries in Zimbabwe leasing out lands informally—Mudimu et al. 2021). This kind of redistribution has occurred in many guises and in many places, historically. These include the many private land titling initiatives, past and present, that were captured by elites in which the poor lost access to and control over land resources, as shown in the huge critical literature on the subject. The Vacant, Fallow, and Virgin (VFV) Land Law in Myanmar was framed as giving land tenure security to ordinary rural villagers, but in reality, the vast majority of land covered in this policy is destined to be reallocated to owners of big capital. Perverse redistribution is a direct outcome, in most cases, of the neoclassical economics ideas of allocative and distributive efficiency in which land ought to be reallocated from what are judged to be inefficient land uses and land users to efficient ones. The result is a net flow of land-based wealth and power from poor working people to-well-to-do farmers and owners of big capital.

Third, *lopsided distribution* occurs when land-based wealth and power are transferred from the state or community, directly or indirectly, by policy or through the open market, to a handful of private or state entities, with the net effect of excluding others while benefiting a few. Many examples of this can be seen in the forest land allocation process in Vietnam from 1994 onwards, whereby forest resources were devolved to communities but then were generally captured by local political elites (Sikor 2006). Moreover, in varying forms and degrees, many land privatization processes (or attempts) in ex-socialist countries also appear to fall into this category (see Spoor 2008 for a global perspective). Finally, twenty-first century land reform contestations in Bolivia can be traced back to the 1953 revolutionary land reform which brought radical land reform to the western part of the country, while the eastern part saw the landed classes and well-connected entrepreneurs gaining control over vast tracts of public lands (Kay and Urioste 2007).

(Re)distributive reforms

Specifying whether a land policy is distributive, redistributive or non-redistributive is a useful analytical reference point to determine the political character of the reform, that is, to answer the fundamental question: who gains, who loses? In this entry we focus on (re)distributive reforms; there is a separate entry in this book for non-redistributive reforms.

Jonathan Fox (1993: 10) offers a useful guide on the distinction between distributive and redistributive reforms:

> Distributive reforms are qualitative changes in the way states allocate public resources to large social groups.... Redistributive reforms are a special case of distributive policies: they change the relative shares between groups. This distinction is important for two principal reasons. First, many apparently redistributive reforms are not, and to call them so implicitly begins with what should be the ultimate outcome of analysis: determining what a social reform actually does, and why. Second, redistribution implies zero-sum action, whereas social programs often are carried out precisely because they avoid clearly taking from one group to give to another. In a context of economic growth, moreover, antipoverty spending may well rise in absolute terms without changing its relative share of the government budget. The label 'redistribution' builds in an assumption about where the resources come from, whereas the notion of distribution limits the focus to who gets what.

The defining principle of *redistributive reform* is redistribution of land-based wealth and power from the monopoly control of either private landed classes or the state to landless and near-landless working poor (poor peasants and rural labourers). It is a 'zero-sum' reform process, although redistribution is a matter of degree, depending on the net loss of landed classes and the net gain of the landless and near-landless poor. The conventional notion of redistributive land reform, as applied only to large private lands, is the most commonly understood example of redistributive land reform (see 'Land reform', this book). However, there are a variety of other policy measures that can change the relative shares of land held by social classes and groups.

The basic defining character of *distributive reform* is that the landless and near-landless working poor are the recipients of land-based wealth and power transferred to them. However, in the process of land distribution, no private entity suffers any loss of land, for example, when the state distributes lands of its own. In many settings, this type of reform entails affirming and protecting pre-existing land access and occupancy by poor peasants or Indigenous peoples whose tenure is not formal and official. It is a 'positive sum' reform process. It does not take resources from one group in society to redistribute to another. In fact, as the quotation from Jonathan Fox above suggests, such policies are often implemented specifically to avoid having to resort to redistributive policies. For example, a piece of land that is officially categorized as public or state forest is actually agroforest land tended and tilled by poor peasants or forest dwellers. A long-term forest land-use rights allocation is issued to the poor peasants or forest dwellers in order to make their pre-existing access to the forest land more formal and secure. This is a distributive reform.

The landed property rights that are (re)distributed can be private, state or community owned. Organizationally, distributed landed property rights can be individual, group or cooperative. The distributive type of reform is generally less controversial or conflictual than the redistributive type, since the focus is on giving to one group while avoiding taking from another. But it would be a serious mistake to assume that all distributive reforms are conflict-free. This is certainly not the case. Many so-called public lands are sites of persistent and heated struggles between various social groups and classes to gain access to and control over the land resource (Franco 2008), especially where there is a perception by some elites that such distributive reforms may erode some of their economic privileges, prestige and opportunities, whether these losses are real or perceived, as in some cases of commercial farmers in southern Africa. As in redistributive types, distributive land policies can take a variety of forms, including conventional land reform, forest devolution, public land resettlement, and so on.

For some kinds of land policies it is not always easy to say whether they are redistributive or distributive. These include land restitution, share tenancy, land tenure reform, land stewardship, Indigenous land rights recognition and labour reform, regardless of whether the policy is applied to private or public land. A label attached to the name of a land policy does not guarantee redistributive content. The key is to establish the degree to which land-based wealth and power are redistributed, and whether there is a change in the relative shares of different classes and groups.

Land restitution is redistributive in character only when the entity that grabbed the land, or the entity that has benefited from taking over the land, lose their control of that land, and when those who previously lost the land regain access and control. In practice, most land restitutions do not fall strictly within this category. In that case, they are likely to be instances of non-redistributive land restitutions (see separate entry, this book).

Share tenancy reform has been a popular land policy, especially in the twentieth century. Often seen as an alternative or an accompaniment to expropriatory redistributive land reform, share tenancy reform means changing the terms of the relationship between the property owner and the tenant. This reform is redistributive in nature if the property owner loses something, which the tenant then gains. To give a hypothetical example: in coconut plantations, the sharing arrangement between a landlord and a tenant-farmer, usually quite informal, might set up a 75/25 split in the harvest in favour of the landlord, with the tenant-farmer taking care of all labour and production costs. Moreover, the landlord can cancel the arrangement and expel the tenant from the land at any time. A law is then passed that bans this kind of informal sharing, shifts the arrangement to a formal one (with contracts, etc.) and fixes the lease rental equivalent to roughly 25% of the harvest during the preceding year; all productivity increases after the new arrangement will accrue only to the tenant-farmer, who also enjoys a security of tenure protected by law. The landlord can no longer expel the tenant from the land without just cause or without compensation implications. This is an example of a redistributive share tenancy reform, as it radically inverts the relative shares of the landlord and tenant and changes the terms of such a relationship. Many such share tenancy reforms were passed in many countries during the past century, mostly as alternatives to or complementary with expropriatory land reforms (see Borras 2007). But there are many other tenancy reforms that are not redistributive in character (see 'Non-redistributive reforms', this book).

The array of state and social forces that emerge for or against land policies broadly depends on what kind of policy is at stake, that is, redistributive or distributive. The former will usually provoke fierce resistance from those who stand to lose wealth and power.

Restitution

> The return of objects that were stolen or lost; payment made for damage or loss
>
> (Cambridge Dictionary n.d.)

One of the most overt forms of injustice and oppression is the forcible expulsion of people from their land. It is not only lands that are taken; a significant part of their lives, time and possibilities in life are also stolen from them, along with the community within which these now displaced peoples used to be embedded. There are tens of millions of internally displaced peoples in the world today. For example, in Myanmar, the combined expulsion of Rohingya people and the displacement of many communities since the military coup in February 2021 has resulted in at least 2 million displaced peoples. The prolonged violence in Colombia has led to the displacement of somewhere between 5 million and 8 million people. The pre-1994 apartheid regime in South Africa forcibly displaced millions from their land. In dealing with such complex social phenomena, and following the Cambridge Dictionary definition of restitution, the questions to be addressed are: what has been stolen or lost? What is the value of the damage and loss? These are basic questions that could help us grapple with what land restitution means in theory and practice.

Let us take a hypothetical example: a community of villagers (call it village A) collapses when the people are forced to abandon their land due to violent conflict and militarization. The villagers become internally displaced people (IDP) living in refugee camps or scattered across distant cities and towns. Companies engaged in extractive activities move into village A and start mining operations in one part of the village; a timber company arrives and clear-cuts the forested areas for timber; a third company establishes a new oil palm plantation. Some ordinary working people who were fleeing their own original village (village B) because of violent conflict end up resettling in the abandoned residential area and farmlands of village A. Some time later, the government passes a law on land restitution that is based on individual claims. Some of the individuals from village A successfully hurdle the administrative obstacles and are awarded individual plots in their original community. These individuals go back to claim their plot only to find that the community—the landscape—within which the original farm plot was embedded has gone, destroyed and replaced by a completely different landscape in which rebuilding the farm plot does not make sense, and is unlikely to succeed. Part of the plots awarded to the individuals are already covered by the oil palm plantation and other parts of the awarded lands are adjacent to the plantation, in areas where soil fertility has been compromised by the heavily chemical-based operation of the plantation. Some of those who were awarded individual plots find that the land allocated to them is contaminated by the tailings of the mines, while others discover that their lands are now occupied by another set of internally displaced people, from village B.

In this hypothetical case, what was destroyed and stolen was not only the farmland for production and a house lot, but an entire web of access to 'land in production and social reproduction' (see separate entry, this book), and the whole community. The integrity of the agroecological zone or landscape has been undermined to the point that it can no longer support the renewal of social life. Thus, restituting an individualized plot of land in a landscape whose integrity has been compromised would result not in the restoration of what was stolen and destroyed, but the effective ratification of what exists as a status quo postexpulsion. It would not be surprising to hear that, in our hypothetical case, some beneficiaries of the restitution programme immediately sold the plots they had been awarded, opting to monetize the restitution award instead. What was destroyed was the community, the landscape, socio-agroecological zone, territory, or homeland; what is being restituted is a farm plot. What was destroyed was the mosaic of land-based social relations; what is being restituted are small, partial fragments of that mosaic. This hypothetical example highlights the strategic importance of maintaining or reconstructing the integrity of a landscape, zone or territory in relation to an individual farm plot, to serve as the material basis for the incessant renewal of life and society. The hypothetical case is, in fact, all too close to some of the real-life experiences of land restitution programmes.

What is to be done? The first step in reframing how we think about restitution is to be clear about what was stolen and destroyed, and what are the things that have value and how much? Since we are talking about 'land restitution', then the first question must be: what is land in social life? In our separate discussion on land in production and social reproduction (this book), we emphasize the importance of treating these two spheres as a connected whole. When we do that, a plot of farmland is just as important as a house lot, just as important as access to a water spring and to the community forest and public space for playgrounds, and so on. Social life cannot be reduced to, understood in the context of, and sustained within just a plot of farmland. A landscape becomes key, including the social structures, institutions and socio-ecological conditions and processes that sustain the incessant renewal of life and community in such a landscape—however full of contradictions and exploitative social relations that landscape might be. It is this life-renewing landscape that was stolen and destroyed. Giving a household a plot of land to restart life within a destroyed landscape just does not make sense. For us, then, land restitution has to be framed as nothing less than landscape restitution, with all the inherent requirements for the reproduction of social life being reconstructed.

There have been two major land restitution programmes in recent times: the South African experience since 1994 and the Colombian case since 2011. In South Africa:

> The restitution of land rights programme has been implemented through five different options: (i) restoration of the actual land from which people were displaced; (ii) provision of alternative land in cases where the original land is not available; (iii) financial compensation for the lost land; (iv) 'alternative relief,' which includes a combination of two or more of the above types. The fifth option involves giving 'priority access to state resources' for land claimants, and this often entails priority access to free housing or infrastructure development provided by the state, although this option is rarely offered nowadays.... [B]y the end of 2015 over 92% of land claims in South Africa had been settled through financial compensation.... The dominance of cash compensation has contributed to the lack of substantive redress and the inability of the land restitution programme to contribute to agrarian change in South Africa.
>
> (Chitonge 2022: 725)

Ten years into the implementation of the land restitution programme in South Africa, an assessment by Cherryl Walker (2006: 87) that showed the outcomes to be less significant relative to the objective need, brought her to the contested meanings of land restitution. Individual compensation for the stolen plot of farmland or house lot or displacement was the dominant model. This was combined with the resolve of the South African government to prioritize national capitalist development, to preserve what existed, namely, white commercial farms, commercial and residential property of the dominant socio-economic class. Thus, land restitution, instead of a social justice instrument to redress past injustice, becomes an issue of 'disturbance compensation'.

In Colombia, decades of violence in the countryside have resulted in widespread displacement.

> This has had a devastating impact on the millions of Colombians who traditionally rely on land for their survival, especially Indigenous, Afro-descendant and peasant farmer communities. It has led to the forced displacement of almost 6 million people – nearly 13% of Colombia's population and one of the highest displacement levels in the world – and the illegal acquisition of around 8 million hectares of land, some 14% of Colombia's territory.
>
> (Amnesty International 2014: 5)

The main method of restitution in Colombia is through individual claims. As of 30 June 2024, under the 'individual route' of the land restitution law, there had been 156,276 applications for land restitution, with 137,519 of these reaching the judicial stage. Among these, 9,312 decisions had been issued and 15,802 applications resolved, involving 13,092 individual properties, covering an area of 254,927 hectares, benefiting 11,088 families. According to the Land Restitution Unit, 11 million hectares have been requested for restitution, with 50% of these cases having begun their process since 2022 (Government of Colombia, 2024).

Recommended reading

Cambridge Dictionary. (n.d.) https://dictionary.cambridge.org/dictionary/english/restitution.

Chitonge, H. (2022). Resettled but not redressed: Land restitution and post-settlement dynamics in South Africa. *Journal of Agrarian Change, 22*(4), 722–739.

Government of Colombia (2024). Estadisticas de Restitucion de Tierras. Bogota. https://www.urt.gov.co/estadisticas-de-restitucion-de-tierras; https://www.urt.gov.co/-/de-11-millones-de-hectáreas-solicitadas-en-restitución-en-13-años-de-gestión-el-50-corresponde-al-gobierno-del-cambio. Accessed 20 August 2024.

S

Sandwich strategy

The outcomes of redistributive land policies are not predetermined by either structural or institutional factors alone or by the actions of state policy elites alone; rather, the political actions and strategies of a wide range of state and societal actors also have a bearing on the outcomes of the reform process.[1] The symbiotic interaction between autonomous societal groups from below and strategically placed state reformists from above provides the most promising strategy to offset antireform forces and manoeuvres. Jonathan Fox (1993) calls this a 'sandwich' strategy.

Redistributive land policies, such as land reform and land restitution, are politically charged because the questions of 'who gains and who loses' from such policies provoke political tension and conflict. Often these policies also try to recast private property relations, which is perhaps the most difficult political challenge in any societal reforms. When such policies truly strike at the heart of what reform is all about, namely, redistribution of land-based wealth and political power; then, they cannot be treated as simple technical issues. Such policies are political, and profoundly conflict-ridden. There is no such thing as a land policy that generates political consensus within the state and in society. Inspired by and building on Fox's concept of interactive state–society approaches in carrying out redistributive reform (Fox 1993), we see the study of the politics of redistributive and distributive land policies as being dominated by two broad and competing perspectives, namely, state-centred and society-centred approaches.

State-centred approach: The key units of analysis of this approach are the state, policy elites (policymakers and managers) and the agencies or organizations responsible for implementing public policies. Exponents of this approach see the state as an institution of governance, autonomous from society. Taking the state as an independent actor and independent variable, scholars often assume that the state is autonomous in making policy choices and in transforming those choices into authoritative actions even when these run counter to the interests of the dominant classes or groups in society (Grindle and Thomas 1989). Many scholars of this approach place a premium on

DOI: 10.4324/9781032658438-14

the administrative design of the policy, believing that such a policy, if carried out by an efficient state organization, has little reason to fail. State-centred approaches often view social actors such as peasant movements and NGOs as necessary complements to the state's reformist efforts, recognizing the practical administrative and fiscal limitations of the state. This realization has led many policy elites to try to form government-sponsored peasant organizations or to reach out to existing ones in order to reshape them within their own parameters by assigning them specific 'supporting roles' in policy implementation. Concerned with efficient policy implementation, policy elites tend to assume that there is a need to avoid conflict; they therefore do not seriously challenge entrenched elites and do not always welcome sharp criticisms from social movements. Finally, founded on the premise of state autonomy, state-centred approaches posit that state intervention can overcome structural and institutional obstacles mounted by landowners by mustering sufficient 'political will' to effect reform, by efficient administrative and technical organization and by ensuring sufficient funding to finance redistributive land reform.

Society-centred approach: Taking social classes and interest group formations in society as their key units of analysis, advocates of a society-centred approach identify social mobilization from below as the key to land reform implementation. These scholars emphasize the inherent structural and institutional obstacles to reform and the 'captivity' of the state to the interests of the dominant social classes; the state then is assumed to have no autonomy. Some stress the influence that social forces exert directly on the state, while others highlight the external constraints they impose. This approach assumes either that the state is monolithic or that any internal differences within the state are direct reflections of societal interests. The activities of the state and policy elites are understood to be dependent variables. Thus, the policy choices and the behaviour of policy elites can be predicted on the basis of an analysis of class and group formations in society or in the international arena. Proponents of this framework tend to argue that proreform forces must pressure the state into implementing land reform. Thus, effective peasant organizations, NGOs, and political movements must necessarily be 'independent' from the state. Moreover, social mobilization from below sets the parameters, extent and location of reforms; state actors only react to such pressures. The relationship between proreform societal actors and the state is necessarily conflict-ridden, and oppositional pressure politics is the most effective way to press for reforms. Focusing the analysis on social classes and class alignments based on a view of the 'prior distribution of power' in society, this approach argues that to overcome the structural and institutional obstacles to land reform, substantial, even if partial, structural, and institutional changes must first occur within the state and in society.

Both of these approaches have limitations. The state-centred approach overemphasizes the autonomy of the state and the capacity of policy elites to overlook the weight of structural–institutional factors, societal groups and international institutions in obstructing redistributive reforms. This approach finds it difficult to explain why good public policies on redistributive reform, implemented by capable administrative organizations and sufficiently funded, can still fail. The society-centred approach, on the other hand, overplays the significance of structural–institutional factors, societal groups and international institutions, neglecting the importance of state and policy elites in conceptualizing and implementing redistributive reforms. This approach cannot easily explain why, in some cases, state actors undertake autonomous actions that run counter to the interests of the dominant classes in society.

By adopting such one-dimensional views of state–society relations, both approaches have difficulty explaining why, in many cases, societal actors attempt to influence and transform state actors, but in the process are themselves transformed and vice versa. On peasant mobilization specifically, society-centred approaches often struggle with the issue of why the actions of strong, independent peasant movements have led in many cases not to sustained land redistribution, but to violent retribution by the state and landlords. At the same time, state-centred approaches cannot fully explain why co-opted peasant organizations, often organized by policy elites as part of the state's extended administrative machinery, usually fail to perform even the 'supporting roles' assigned to them. A third approach offers a different perspective.

Interactive state–society approach: The process of implementing public policy entails an interaction between state actors and societal actors, despite the latter's usual claim that they are independent groups. Two concepts must be clarified here: independence and autonomy. Independence is largely seen as an absolute 'either/or' question: groups are either co-opted by, or are independent from, the state. To be independent means that the internal dynamics of societal organizations are (and can be) insulated from any form of state interference or influence. Meanwhile, autonomy is inherently a matter of degree and refers here to the amount of state intervention in the societal actor's internal decision-making (Fox 1993: 28). Independent social movement organizations can only go as far as putting issues onto the state's agenda, but they are generally unable to directly influence policy outcomes without close interaction with state actors. Co-opted organizations do not make a significant impact since they are basically administrative adjuncts of the state and can rarely go beyond what the state defines as parameters of action. Autonomous organizations have more potential. While they are able to penetrate the state from top to bottom and to influence it from within, they can also pull out of such interaction when disengagement is necessary and preserve themselves when the windows of opportunity close, still retaining some degree of strength from previous interactions with the state which can be utilized for the next reformist opening.

In some cases, state reformists and pro-reform social groups exist without interacting with each other. In such situations, political opportunities are not harnessed. In other cases, they do interact but instead of supporting, they undermine each other: pro-reform forces are then generally weakened and prospects for land reform implementation are remote. The most promising situation is when the two streams of pro-reform forces from above and from below interact in pursuit of the common goal of implementing land reform, despite differences between them in agendas and motivations. This positive interaction does not necessarily entail explicit coalitions between state and societal actors. Parallel initiatives of the state and societal actors (who may even consider themselves adversaries) toward a common aim also form objective alliances. In short, each must pressure the other to give in, but they share a broader interest in each other's gaining strength. The different motivations underlying the actions of the state and societal actors are responsible for the inherent potential for conflict in the relations between objectively allied state reformists and societal actors (see Fox 1993: 21–32).

Achieving a symbiotic interaction between pro-reform state and societal actors does not automatically lead to land reform implementation. The pro-reform forces have to surmount various anti-reform obstacles. There are a number of ways in which the anti-reform coalition can be weakened, and here, the issue of political opportunity is relevant. Tarrow (1994) identifies four important political opportunities: access to power, shifting alignments, availability of influential elites, and cleavage within and among elites. The availability of all or some of these opportunities can create possibilities that even the weak and 'disorganized' actors can take advantage of: conversely, even the strong grow weak. Thus, pro-land reform forces can overcome antireform obstacles when their ranks remain solid and persistent, while the landlords may either fail to muster sufficient state allies, may be abandoned by state allies, may face a split from other elites or may simply be overpowered by the composite force of the pro-reform actors.

Recommended reading

Borras, S. M. (2007). *Pro-poor land policy: A critique*. Ottawa: University of Ottawa Press.

Fox, J. (1993). *The politics of food in Mexico: State power and social mobilization*. Ithaca, NY: Cornell University Press, especially Chapter 2.

Scarcity

A basic condition for land boom or land rushes to occur is the perception of scarcity of particular types of land. Land will not be as exciting and important topic for conversation and policy and political debates if all types of land are available for everyone who needed them, wherever and whenever these are required. That land in the context of production/social reproduction connected whole is scarce is almost an inherent feature of land, otherwise it will be a free resource. Land scarcity is a central concept in conversations about agrarian transformations. But how scarcity is understood and socially constructed is politically contested, putting the making of scarcity and its definition inherently in the plural form.

Scoones and colleagues offered a useful typology that is an effective heuristic tool to help us make sense of and critically analyse the concept of land scarcity (Scoones et al. 2019). They explain that there are three types of scarcity: absolute, relative, and political. Absolute scarcity is understood as 'physical, real and inescapable' and thus is a problem of a finite supply of a resource. This is a (neo)Malthusian perspective. This perspective includes views locating agricultural land scarcity or the exhaustion of the land frontier relative to the need to protect remaining forests, biodiversity and other spaces that are important for conservation and sustaining the biophysical carrying capacity of the planet.

Relative scarcity recognizes the finite supply of a resource but assumes that through technological advancement, as well as institutional intervention, such as through price signals, such limits can be overcome. This view assumes that technological and institutional innovation and intervention could be accomplished through various possibilities of substitution of the scarce resource. For example, limited supply and thus high cost of domestic labour can be addressed by taking in migrant workers, or, through the mechanization of work, as for example, the scarcity of sugarcane cutters could be remedied by mechanizing sugarcane harvesting. The same substitution logic applies in terms of other resources, such as food or energy. Substitution can also be complemented by resource recycling. In terms of land, the narrative from this perspective is focused on land and labour productivity increases, that is, how can more commodities be produced in the same limited area of land. This is where the popular topic of 'yield gap' is located in which there is a normative assumption about the land productivity level of which commodities (Deininger and Byerlee 2011).

The third ideal type in the typology by Scoones et al. is political scarcity. This perspective is based on the assumption that resource scarcity, such as land scarcity, is located within the logic of capital under a global situation of generalized commodification. Scarcity therefore is not seen as something 'natural' but is rather politically constructed (Scoones et al. 2019). This view does not altogether dismiss some issues related to productivity as has been raised in the second ideal type but sees such issues as something that can be addressed through social and political transformation of the land-based social structures.

Of the three ideal types, our argument in this book identifies with the political scarcity view. In Colombia, many of the signatories of the 2016 Peace Agreement, or what is also called 'ex-FARC combatants', have been disarmed, demobilized and resettled in temporary housing sites often in remote rural areas. The conditions are not good. There are no job possibilities nearby, no land to farm, and the land in which the temporary houses are installed is not at all secured. In short, many of them live like squatters in rural slums. There is no land for productive and social reproductive activities for the Peace Agreement signatories (Rojas 2023). Yet, if you are a prospective cattle rancher or an agribusiness or mining investor, there is abundant supply of land that you can lease or buy. It is thus a situation of land abundance for those who have money and power, land scarcity for the working people without political clout. This is an example of what is meant by land scarcity from a political perspective.

The only way to address political scarcity therefore is to restructure social relations around land for production and social reproduction. But the idea of 'land redistribution' itself is a politically contested concept, as we have discussed elsewhere in this book (see, for example, 'land reform' versus 'market-assisted land reform', this book). For us, following a political understanding of scarcity, a fundamental element required in effectively addressing land scarcity is the notion of 'minimum access/size ceiling' (this book).

Recommended reading

Scoones, I., Smalley, R., Hall, R., & Tsikata, D. (2019). Narratives of scarcity: Framing the global land rush. *Geoforum, 101*, 231–241.

Scholar-activism

We locate our work on the politics of land within the tradition of scholar-activism and in particular the kind of scholar-activism in the fields and fronts of agrarian, food, environmental and climate studies and struggles.[2] Thus, what concepts on land politics we deem relevant and important and how we define such concepts also take a scholar-activist perspective.

Scholar-activism is a way of working that tries to change society by combining the best features of radical academic and political activist traditions, despite the many contradictions and challenges that this entails. It is not a politically neutral scholarship. In fact, it is politically biased and has its own normative assumptions. Its intellectual, political, and moral compass is the social justice struggle for a world that is more just, fairer and kinder. It necessarily takes a bias in favour of the exploited and oppressed classes and social groups. Scholars in social sciences who claim to be neutral are often confusing neutrality with rigour. Social science scholarship can be rigorous, but it is difficult to think of how it can be politically neutral. We choose to study different social problems because we judge them to be problems. How we define problems entails some kind of social values, and addressing such problems requires normative judgements. For example, grabbing land from ordinary villagers for the purpose of establishing large-scale monoculture plantations may be seen as a welcome positive investment, and thus not a problem, while scholar-activism that takes the side of the marginalized and oppressed social groups will see it as an urgent problem. How research questions will be framed from the two perspectives will be radically different.

The methods can themselves be neutral, although the choices of what methods to use are influenced by the way we have defined and framed the research, which is, in turn, normative. Our choices of analytic tools entail assumptions that influence not only the methods we use and the empirical data we gather, but also what causal relationships between factors we expect to find. So, the framing of research is a normative or activist act.

Scholar-activist workstreams are not the only ones trying to change scholarship and society; indeed, they probably constitute little more than a trickle among the wider communities of academics and activists. There are many radical thinkers who spawn ideas that are important for political struggles but are not committed to particular political projects or movements or the deployment of specific research methods or research protocols that are officially sanctioned by movements. There are also activists who are completely preoccupied by practical politics 'in the trenches', whose work is sufficiently informed by appropriate theories and concepts; they are found in a range of sectors, from environmental movements to trade unions. Similarly, scholar-activists' particular work and contributions are important to political work but, on balance, constitute only a small portion of the latter.

There are diverse and multiple spheres of knowledge production. But for the kind of agrarian scholar-activist tradition that we identify with, there are three key hubs, namely, academia, social movements and autonomous research institutions. The knowledge that each of the three produces is distinct, and for us, is of equal value. We do not see it from a hierarchical perspective where one type of knowledge (often, the academic science) is deemed as superior over the others. Rather, we see it from a horizontal perspective where the three hubs produce distinct knowledges of equal relevance and value.

Recommended reading

Edelman, M. (2009). Synergies and tensions between rural social movements and professional researchers. *The Journal of Peasant Studies, 36*(1), 245–265.

Scope (land grabbing, land rush)

One of the contentious issues in the literature on land grabbing which has emerged since 2008—referred to by Oya (2013) as the 'literature rush'—is the question of 'scope'. What do we count, how do we count and why do we even have to count? These are not trivial questions. In fact, they are highly debated topics (Anseeuw et al. 2013; Edelman 2013; Oya 2013; Scoones et al. 2013). This is important because it has implications for what it is that we are studying in the first place. Therefore, 'what we count' is closely linked to how we understand concepts. In our understanding of the idea of 'scope', it is important to disaggregate and differentiate the two concepts of land grabbing and land rush, as we have done in this book (see separate entries, this book).

The contentious discussion about scope occurs within the parameters of the definition of land grabbing and the many other terms that are used interchangeably—large-scale land acquisition, large-scale land investment, large-scale land transaction, or land deals—in other words, within a demarcated plot of land at a given period involving a land grabber and the land grabbed. The biggest disagreements stem from what data are collected and deposited in a database. The best known and most widely used database in this case is the Land Matrix (Anseeuw et al. 2013). There are several concerns about the use of this dataset, or for that matter, any other large datasets on land grabs. First, the definition of land grabbing is not universally shared, and in building the database, things are bound to be missed. In the early years of the land grab debate, the Land Matrix excluded land grabs in the Global North as well as big conservation projects, an issue which has been flagged by critics. Second, information entered in databases freezes extremely dynamic and evolving social phenomena, resulting in flawed data about the status of land deals. Third, the cut-off lines in terms of the size of a deal—sometimes set at 1,000 hectares, sometimes at 500 hectares, and still other times at 200 hectares—are too land-centric and miss a vital element of land grabbing which is the scale of capital in the operation (Edelman 2013). Oya's critique in relation to large databases, especially in politically fluid phenomena like land deals, is pointed: 'garbage in, garbage out' (Oya 2013). One general conclusion from these critiques is that databases such as the Land Matrix overestimate the scope of land grabbing. The institutional design of the Land Matrix has been progressively improved over time to address these early criticisms. (For more on this series of improvements and internal re-reading, see Anseeuw et al. 2013; Nolte et al. 2016; Lay et al. 2021.)

Within a few years of the global land grab that started in 2008, it became clear that many of the mega land investment projects that had been so prominently announced and extravagantly launched had been abandoned or cancelled, prompting some observers to call land grabbing a hype (Kaag and Zoomers 2014). From then on, the debate about scope revolved mainly around 'operational' and 'non-operational' or 'failed' large-scale land acquisitions (see separate entries, this book). The main discussion point was whether or not the scope of the land grab was significant, given the many failed land deals. The general tendency was to dismiss it as less significant than the hype had led us to believe.

But scope is not a simple matter of counting. It is deeply conceptual, as Edelman (2013) reminds us. In discussions on scope, there are three major categories around which the academic literature revolves, namely, (i) concluded, operational land deals, (ii) concluded, non-operational/failed land deals and (iii) land rush. These three are distinct from each other, although they are overlapping concepts.

Operational land deals are defined in Land Matrix terms as concluded land deals at various stages of capitalist enterprise development, that is, some are well advanced in their business operations and others are just about to begin, with the rest somewhere in between. These are mostly corporate land deals and large scale. Going back to earlier debates, this category was supposed to represent the 'clean' dataset and the only information that should be used when studying this social phenomenon.

Non-operational land deals are those deals that were concluded but for various reasons they were withdrawn, cancelled, stalled indefinitely, scaled back. This can include cases that were publicly launched through a memorandum of understanding (MOU) between investors and host governments or communities. Good examples of this category are the numerous mega land investment projects by owners of big capital from China in the Philippines, announced publicly after the signing of MOUs between the two governments and representatives of the corporate sector, but most of which never took off. This category also includes land deals in which the capitalist enterprise started but was cancelled at a later stage. Iconic examples include the Procana case involving 30,000 hectares of land in Mozambique that halted its sugarcane plantation operation after just a few years. Another example is the Karaturi company, which managed to get 100,000 hectares of land in Gambella, Ethiopia, clear the land, and begin operations, only to stop a few years later. In Cambodia, Pheapimex company acquired a concession of around 320,000 hectares, but approximately half of this was later retrieved by the government and returned to villagers. The remaining half is still in the hands of investors who pursue sporadic bursts of capitalist enterprise on the land, most of which have never really amounted to anything serious in terms of a business. These three examples are lumped together in the catch-all phrases of non-operational or failed land deals although the majority of the grabbed lands were not returned to the villagers.

The issue of scope starts to get muddled in this category of non-operational and failed land deals. In the three examples cited above, what failed, at least for the time being, were the planned capitalist enterprises; the land grabs themselves were successfully accomplished. Thus, not including these cases in the scope of land grabbing is a mistake. To use this category to dismiss land grabs as merely a hype and suggest that the scope of land grabbing is much smaller than earlier projected is a serious error.

Furthermore, if we take on board the concept of land rush (see separate entry, this book) to complement our concept of land grabbing, then the scope of land grabbing may actually be much higher. The scope of land rush will be different from the scope of land grabbing. Our definition of land rush speaks of a frenzied, hyperbolic, spectacularized mad scramble for land. Where a full-scale land rush happens, the official purveyors of land and investment prospecting are joined by a multitude of land entrepreneurs, aspiring farmers, land brokers, scammers, swindlers, thugs, shady investors from the illicit economy sector (illegal drug syndicates, armed paramilitary groups, and so on). They engage in land transactions, many of them on an individual level, by stealth and informally; many of these transactions are small-scale

(5 or 10 hectares, for example), and often occur out of the spotlight. These are pin prick land grabs (see separate entry, this book). Separately, these small and scattered individual land transactions do not appear significant. They would not arouse the interest of the media, or the investment and planning ministry, or databanks such as the Land Matrix. But when aggregated they amount to something much bigger, and together they can be just as extensive as land deals by high-profile corporate investors (Borras et al. 2024). Pin prick land grabs are less susceptible to some of the problems faced by large-scale land investments, such as waiting for investment funds after making hyperbolic claims about a newly acquired large tract of land (financial investments which do not always materialize, as in the case of Procana in Mozambique).

The three categories of operational land deals, non-operational land deals and pin prick land grabs are, in the context of a land rush, co-constitutive. If we combine these three categories, then it is clear that the Land Matrix database does not over-estimate the scope of land grabbing. In fact, the dataset is an under-estimate of the real scope of land grabbing. We reached our conclusion on pin prick land grabs by grappling with the concept of land rush as something overlapping with but distinct from land grabbing. When addressing the issue of scope, it is therefore important to include all three categories. But we also need to go beyond these. We define land rush as a speculative, spectacular, hyperbolic and convulsive scramble for land: this also includes those announcements by PR firms that are obviously aimed mainly at drum-beating to arouse the interest of possible investors, such as the massive promotional hyperbole indulged in by the Colombian government and the corporate world declaring Altillanura as the 'new cerrado', the 'new El Dorado' (Arango 2024); the millions of hectares of land implicated in the Prosavana project of the Mozambican government; the dozens of mega land projects in the Philippines announced by owners of big capital from China and the Chinese government; the 1 million hectare Kuok company–San Miguel land deal with the Philippine government. All of these should be counted and accounted for as the scope of the land rush. Fake news stories in mainstream media and in social media about land investments and investors need to be tracked and included in accounting for the scope of the land rush. Only by having a good sense of the scope of the spectacular claims during the land rush can we begin to understand the character of the current conjuncture in global capitalism and the part played by land in its aspiration for continuous reproduction of capital, for the never-ending remaking of frontiers of capital. If we are tracking demarcated plots of land acquired for land investments, whether operational or non-operational, counting them and aggregating them in order to determine the scope of land grabbing, then the kinds of data and information we have just described will have to be screened out. They constitute what Oya calls 'garbage'. But if we want to understand the character and dynamics of the land rush, then the concept of 'scope' has to be radically revised.

Recommended reading

Edelman, M. (2013). Messy hectares: Questions about the epistemology of land grabbing data. *Journal of Peasant Studies, 40*(3), 485–501.
Oya, C. (2013). Methodological reflections on 'land grab' databases and the 'land grab' literature 'rush'. *Journal of Peasant Studies, 40*(3), 503–520.

Sharing

Land sharing and land sparing are two concepts associated with land policy questions related to conservation. The main issue of interest is the relations between land for production and land for conservation. *Land sharing* does not separate land for production from land for conservation, as the concept of 'land sparing' (see separate entry, this book) does. Instead, it aspires to the integration of land for conservation and production. In contemporary narratives, land sparing and land sharing have to be understood in interrelation. As Fischer et al. (2014: 149) explain:

> In a land sparing strategy, some land is set aside for conservation while other land is used intensively to produce agricultural commodities; in a land sharing strategy, less land is set aside specifically for conservation, but less intensive production techniques are used to maintain some biodiversity throughout agricultural land.

Dressler et al. (2016: 76) argue that:

> While recent studies support intensification and zoning as the basis for land sparing, we argue that land sharing in multifunctional landscapes—protecting less land but farming the remainder in agro-ecologically diverse, sustainable ways… is better suited to the complex reality of poor, resource reliant uplanders. This is because poor uplanders' livelihood security depends on a diversity of resources from varied ecosystem services that are sustained through complimentary resource uses that also support forest conservation.

Elsewhere in this book (see 'Land sparing') we illustrate the broader political economy background of the land sparing strategy through a detailed hypothetical case. There we explain that mainstream economics views particular types of production systems, especially the mobile, subsistence-oriented systems such as swidden farming, small-scale mobile pastoralism and Indigenous community production, as 'economically inefficient' or 'ecologically destructive' or both. From this narrative comes the justification for the state to take control of these types of lands, often combining land titling programmes that are at the heart of the sedentarization campaigns targeting these kinds of rural communities with repurposing land use and reallocating land to agribusiness land investors or big conservation schemes, or both. This is the context for land sparing approaches.

Land sharing is pitched differently: its aim is the integration, rather than separation, of production and conservation in an area of land. The general idea of land sharing is that gains from conservation can be realized in and through production, as long as these are achieved in agroecologically sound and sustainable ways, as described by Dressler et al. (2016).

However, while subsistence-oriented production may be able to feed the households engaged in such production, it may not be able to produce a marketable agricultural surplus to feed the rural and urban populations that do not produce food themselves. Thus, production and productivity increases remain a challenge, if these have to be achieved without using technologies that destroy biodiversity and harm the environment more generally, such as the extensive use of fossil-based inputs, pesticides and other chemicals. The transformation of production towards an agroecological method is a promising path, and there is evidence that productivity increases can be realized through this method (Rosset and Altieri 2017; Akram-Lodhi 2021; van der Ploeg 2021). The question that remains is how fast agroecological scaling-up can be achieved worldwide. Nevertheless, the concept of land sharing—integrating production and conservation in a landscape—offers a way of fusing together environmental and climate justice with agrarian justice, or a form of 'agrarian climate justice' (Calmon et al. 2021; Sekine 2021; Yaşın 2022).

Recommended reading

Dressler, W., de Koning, J., Montefrio, M., & Firn, J. (2016). Land sharing not sparing in the 'green economy': The role of livelihood bricolage in conservation and development in the Philippines. *Geoforum*, *76*, 75–89.

Social function of property

One of the most politically consequential peasant movements of the late twentieth century and early twenty-first century is the Landless Movement (MST) of Brazil. Formed in 1985, it has become renowned globally for framing and pursuing a political strategy of land occupation to demand land redistribution (Wolford 2010). Its political mobilization is strategically anchored on a constitutional and legal opening that provides political opportunity for land claim-making from below. This is the legal provision for the 'social function of property', which is enshrined in the Brazilian Constitution, and provided an operational guide for the 1988 land reform law. So, what is the social function of land property?

The concept of the social function of property originated with the French legal scholar, Léon Duguit (1859–1928) whose main aim was to promote social solidarity. Duguit suggested that private property has a social function

(Ondetti 2016); more specifically, and in relation to land, the idea of a social function of land property is that 'the right of private ownership includes an obligation to use land in ways that benefit society as a whole' (Ondetti 2016: 29), which is interpreted throughout Latin America to mean 'the positive responsibility to use land for productive purposes' (Ondetti 2016: 29). Duguit's original idea, according to Ondetti, was not to use this concept for the purpose of effecting land redistribution, but to encourage the full productive utilization of land.

The scope as well as the limitations of this concept in terms of actual land reform policy and politics in Brazil are summarized by Ondetti (Ondetti 2016: 29) as follows:

> At the same time, however, the constitution states flatly that 'productive' property cannot be expropriated. For many analysts this provision essentially nullifies the labor and environmental aspects of the social function, since it implies that productive farmland cannot be seized no matter what environmental or labor-related infractions the owner may be guilty of. Although agrarian reform activists have sought to pressure authorities to implement these elements of the social function, in practice the ban on expropriating productive land has prevailed.

Despite the inherent limitations of the concept of the social function of land, and the way it has been translated into laws, it has inspired and provided political opportunities for mass land claim-making by the landless. Ondetti's assessment of how this was interpreted and implemented and with what results are captured in the paragraph below:

> The efforts of the last two decades have significantly increased the reform sector in the Brazilian countryside. As of the end of 2013, the state had expropriated 7005 properties with an area of 30,466,875 ha. Using data from the... [2006] agricultural census, the latter figure represents 9.2% of Brazil's farmland. *All, or virtually all, of these expropriations have been based on productivity*. For reasons discussed in Section 3, the constitutional provisions regarding the labor and the environment aspects of the social function have been virtually a dead letter. [O]fficial records indicate that close to a million families have been granted land. However, because some families have left their plots, and because there may have been some padding of figures over the years, the actual number of settler families is undoubtedly smaller.... One study based on the 2006 farm census put it slightly below 600,000, which represented close to 12% of all agricultural establishments at the time.
>
> (Ondetti 2016: 32, emphasis in original)

It is useful to read this in comparative perspective with the concept of 'minimum access/size ceiling' (see separate entry, this book).

Recommended reading

Ondetti, G. (2016). The social function of property, land rights and social welfare in Brazil. *Land Use Policy, 50*, 29–37.

Sparing

Land sparing is a strategy to separate land for production from land for conservation, with the idea of maintaining or even increasing production output through intensive production systems for increased productivity in a smaller area of land, while liberating the rest of the land from production. The latter area is then demarcated for conservation, such as reforestation, biodiversity protection and the like (Fischer et al. 2014). As Phalan et al. (2011: 1289) explain:

> land sparing... consists of separating land for conservation from land for crops, with high-yield farming facilitating the protection of remaining natural habitats from agricultural expansion.... Achieving land sparing is fundamental to reducing emissions from deforestation and forest degradation (REDD) and requires the sustainable intensification of agriculture.

Dressler et al. (2016: 76) notes that: 'With few exceptions, most land sparing approaches promote spatially constrained, sedentary agriculture and conserving "high value" ecosystems in fixed zones'. The concept of 'land sparing'—almost always paired with the contrasting idea of 'land sharing' (see separate entry, this book), has sparked controversy and debate since the beginning of the twenty-first century (Phalan et al. 2011).

Let us illustrate what this means through a hypothetical example. There is a landscape that is predominantly characterized by swidden agriculture (also loosely referred to as 'mobile agriculture' or 'shifting cultivation'). There are 50 households, all engaged in this kind of farming. Each household cultivates an average of 5 hectares of land, planted to rainfed maize, rice, sweet potatoes, cassava, plantain and other vegetables such as pumpkin. After two years, the household abandons the cultivated area and opens a new 5-hectare plot, using fire to clear the bushes and small trees and leaves the old plot fallow. After moving five times, they go back to the first plot, now fallow land. It is a thus a 10-year rotation. A household needs 25 hectares of land in total (5 hectares each cultivation lasting 2 years, 5× shifting = 25 hectares). With 50 households practising this land-extensive, shifting cultivation under the terms explained here, the total area under cultivation within a 10-year rotational basis is 1,250 hectares (25 hectares × 50 households). There is no need for external chemical fertilizer for one or two years; beyond that, soil fertility declines, hence the need to open a new clearing. This farming system is largely orientated towards use value and subsistence; when exchange value is produced it is usually aimed at generating cash to purchase other daily needs. Sometimes this is the only source of use value and (some) income produced by the household, but in many other cases, it is supplementary to or is supplemented by paddy cultivation in another field or animal keeping in another area, or wage work elsewhere, or a combination of all of these. This is classic shifting cultivation. There are variations of this in terms of average cultivated land area, length of cultivation in each site, but this hypothetical example illustrates fairly faithfully what we see from Colombia's *chagra* to the Guatemalan *milpa* to Myanmar's *taungya*. The land tenure is usually held under customary arrangement as commons (although often claimed at the same time by the government as state land).

There are two long-standing mainstream narratives that disparage this type of production system. On the one hand, it is seen 'economically inefficient' in terms of land use and land users. When land is seen as a scarce resource, then this type of land use is considered wasteful. The idea of a 'yield gap'—the potential of a given land for maximum production and productivity versus the actually existing level of production and productivity—is used to show how under-utilized this scare resource is. Economic efficiency is often measured in terms of monetary value. For instance, the monetary equivalence of all harvested maize, cassava, sweet potatoes, pumpkins in a 5-hectare plot, versus the potential value of 5 hectares of rubber plantation. Almost always, subsistence-oriented or supplementary livelihood shifting cultivation falls below the much higher level of monetary income under a commercialized, intensive way of farming: sedentary, using farm machines and chemicals and fertilizers to produce exponentially more value from the same size of land. On the other hand, because of the use of fire in clearing land, shifting cultivation is often seen as 'ecologically destructive'. In many countries, shifting cultivation has been criminalized. Separately, these two narratives have justified cycles of state attempts at taking over these lands and reallocating them for various commercial purposes. But since the beginning of the twenty first century, these two narratives have often been deployed together, and thus have become even more potent in terms of taking lands away from peasants, Indigenous communities and mobile pastoralists.

It is in this context that, during the same period, the idea of land sparing has become extremely popular. In short, land sparing is no more than the two narratives above being combined to argue for the reallocation and repurposing of land. The usual line of argument and policy starts with a programme to demarcate and provide private land titles to shifting cultivators (or to mobile pastoralists, or a fixed resettlement area for Indigenous communities). In our hypothetical case, this would be 5 hectares of freehold individual private land property for each of the 50 households. This is pitched as a land tenure securitization measure. It is seductive. There are at least two logical consequences of this: conversion of production from a mobile to a sedentary method, and the end of access by the 50 households to the spaces which they used previously to rotate their production. So, if 5 hectares for each household is privately titled, there would be 250 hectares of land that would have formal individual land titles. The question is: what happens to the 1,000 hectares of land not included in the land titling programme, and thus spared from further cultivation by the 50 households? This much bigger area is then taken over by the state. Where this happens, this bigger area is either reallocated by the state to large-scale commercial agribusiness investment, or conservation programmes, or both.

One of the claimed objectives of land sparing is to help increase the production and productivity level of our hypothetical 50 households in their now fully titled land for production, often with the use of modern technology (farm machines, chemicals, commercial fertilizers and certified seeds), or to promote contract farming schemes with agrarian capitalists. The other claimed objective is to consolidate the area of conservation, including commodified conservation schemes such as REDD+ that often comes in variations of 'fortress conservation' (Brockington 2002).

Our hypothetical case reflects a common reality in many regions of the world today. It is true in the case of swidden farmers, but also among Indigenous communities and small-scale mobile pastoralists. Their community lands are often the primary targets for land sparing.

Recommended reading

Fischer, J., Abson, D. J., Butsic, V., Chappell, M. J., Ekroos, J., Hanspach, J., Kuemmerle, T., Smith, H. G., & von Wehrden, H. (2014). Land sparing versus land sharing: Moving forward. *Conservation Letters*, *7*(3), 149–157.

Spectacle

For a land rush to emerge, the build-up towards it must include the elements of hyperbolic claims, wild speculation and great spectacle. The global conjuncture around 2008 provided the material basis for these elements to come together. During this period, various aspects of capitalist crisis converged: food price spike, fuel and energy crisis, environmental and climate crisis and financial crisis, all linked in some way to the overall crisis of overaccumulation of capital in the global centres of capitalism. The purported solutions to the crises are in turn framed as great opportunities to remake the frontiers of expanded reproduction of capital: finance capital seeking new lucrative investment portfolios, transnational companies seeking cheaper sources of raw materials and labour as well as new markets, and entrepreneurs in newly emerging green capitalism (businesses in renewable energy, carbon trading and so on).

In the narrative that ensued, 'land' was to play a critical role in all these so-called solutions that are, in turn, new frontiers of capitalist accumulation. The neoclassical economics assumption that partly triggered and justified the subsequent land rush was that there is plenty of land in the world that is vacant and unused, or being under-utilized by economically inefficient land users (poor peasants, small-scale pastoralists and forest dwellers) (Deininger and Byerlee 2011). There was a preoccupation with the potential of these lands to produce commodities that could address the crises, and with the so called 'yield gap' (Deininger 2011) between this potential and the actual performance of these lands in reality. It is possible that this view was independently developed and deployed by the corporate sector and state elites who wanted to cash in on the lands in their national territories (aside from neoclassical economists). It certainly became part of the emerging narrative, and has been used to justify large-scale land investments (Deininger and Byerlee 2011). But no one really knew how much suitable land or corporate investment were available, which led to hyperbolic claims, promotional slogans and advertisements such as 'like gold with yield' (Fairbairn 2014). This set the global stage for speculation and spectacularization.

A fundamental element in a land rush is what Tsing (2000: 118) calls an 'economy of appearances', that is, 'the self-conscious making of a spectacle [that] is a necessary aid to gathering investment funds.... It is a regular feature of the search for financial capital'. She elaborates:

> In speculative enterprises, profit must be imagined before it can be extracted; the possibility of economic performance must be conjured like a spirit to draw an audience of potential investors. The more spectacular the conjuring, the more possible an investment frenzy.
>
> (2000: 118)

Moreover, prospecting is a two-way process, with effort emanating from both investors and hosts: 'Nor are companies alone in the conjuring business in these times. In order to attract companies, countries, regions and towns must dramatize their potential as places for investment. Dramatic performance is the prerequisite of their economic performance' (2000: 118). Tsing thus advances the notion of 'spectacular accumulation', which

> occurs when investors speculate on a product that may or may not exist. Investors are looking for the appearance of success. They cannot afford to find out if the product is solid; by then their chances for profit will be gone.
>
> (2000: 141–142)

Using Tsing's concept to look at the land rushes from around 2008 onwards, we see that the conjuring, spectacle and frenzy were far greater than the actual, localized gold rush that Tsing was studying in Indonesia in the 1990s. The land rushes represented a feverish global convergence of old and new actors and forces, driven by hyperbolic projections and perceptions of potential windfalls, and resulting in a convulsion of the existing order amidst fundamental contradictions, uncontrollable urges and insurgent attempts to rapidly recast or reinterpret the institutional rules in the belief that this was necessary to realize the promised windfalls.

We contend that, in land rushes, the number of actors drawn in by the seduction of a potential windfall ends up far exceeding the likely optimum number of investors, and the amount of land that is implicated far exceeds the amount that is realistically needed. At the beginning of the process, no one knows exactly how many investors and how much land are required to reach the optimum point of supply and demand (of commodity and money); this creates an atmosphere of extreme competitiveness in terms of speed and timing, which in turn creates an incentive for competitors to dodge the rules in order to 'get ahead', like the 'sooners' who surreptitiously moved to stake their claim on a prime plot before the official start of the homesteading run during the Oklahoma land rush of the late nineteenth century. Contributing to this febrile, 'get ahead' atmosphere is the fact that most land rushes have elements of extractivism, which implies the potential exhaustion of what is being extracted, not just in terms of land area, but also of water, subsoil minerals, timber and so on, especially in the era of global extractivism (Ye et al. 2020; Alonso-Fradejas 2021; McKay et al. 2021).

A final note: while spectacle is a fundamental element in a land rush, it cannot by itself make or trigger a land rush. The other phenomenon in everyday capitalism related to land accumulation is land grabbing. This also often contains the element of spectacle even when it occurs outside a land rush. It is when a spectacle and a global conjuncture such as that discussed above meet a particular kind of prevailing land regime—the relatively stable pattern of institutional procedures on land use, access, claims, and so on—that a possibility of land rush emerges (see 'Land rush', this book).

Recommended reading

Gutiérrez, J. (2023). Staging the New City: Urban spectacles and the ecological origins of Nayib Bukele's authoritarian populism. *City & Society, 35*(3), 141–155.

Tsing, A. (2000). Inside the economy of appearances. *Public Culture, 12*(1), 115–144.

Notes

1 This entry draws from Borras (2007).

2 This entry draws on Borras and Franco (2023).

T

Territorialization

The meaning of land is not only land as a means of production, for example, farmland or grazing land. Land also means territory (see 'Land in production/ social reproduction', this book). In social sciences, the term 'territory' is commonly associated with the central state's permanent task of demarcating its national boundary with the goal of exercising sovereignty and authority over such a territory. In Critical Agrarian Studies, this common meaning accorded to the term is relevant and important, but it is not the only relevant interpretation. Scholars who have been working on agrarian and environmental issues, and the politics of natural resources more broadly, such as Vandergeest and Peluso (1995), Peluso and Lund (2011) and Shattuck and Peluso (2021), have developed the concepts of 'territory' and 'territorialization' in ways that speak more systematically to scholars and activists in Critical Agrarian Studies. In general, they define territorialization (the process of making territory) as the central state-driven political processes and institutions within one's internal boundaries, with the main objective of governing the politics of agrarian and environmental resources: who gets which land and nature, how, how much, through what mechanisms, why, and for what purposes? (see Shattuck and Peluso 2021).

For Vandergeest and Peluso (1995: 388):

> Territorialization is about excluding or including people within particular geographic boundaries, and about controlling what people do and their access to natural resources within those boundaries. As Sack [1986] outlines, territoriality involves classification by area, theoretically eliminating or altering the need to regulate specific resources or individuals within a territorial zone. Territoriality also involves the communication of both the territorial boundaries and the restrictions on activities within the territory. Forms of communication range from locally recognized markers (a tree, stream, rice bund) to mapped points locating boundaries according to the Global

DOI: 10.4324/9781032658438-15

> Positioning System – a global spatial grid. Registered land titles and maps based on cadastral surveys are the key texts that provide for clear communication of property rights to and among state authorities. Property rights thus become textually mediated.

Territorialization is thus both the classic, state-driven process and a non-state process. For the state-driven process of making a territory, or territorialization, as explained in the extended quote above, is often combined with the use of economic relations or the use or threat of use of the state's coercive apparatus—police, military, courts, and prison—to enforce authority over a claimed territory. These are used to facilitate private capital accumulation in broadly capitalist settings, to collect taxes and fulfil other tasks of the state. Thus, territory is about authority and, following the argument of Rasmussen and Lund (2018: 389): 'territorialization establishes authority (rather than "those with authority can territorialize")'.

When the Colombian government introduced key categories in land allocation, namely, Indigenous communities' settlement lands (*resguardos*), Afro-Colombians' community land rights, campesinos' rights to *baldios* (public lands), *zonas de reservas campesinas* (peasants' reserve zones)—and established official rules and procedures about how to operationalize and maintain them, formally and informally—this was an act of making territories, or indeed, a process of territorialization. But so too was the Colombian government's declaration that some public lands are 'wasteland' and should be allocated for large-scale land investments, as exemplified by its campaign from around 2008 onwards in Altillanura (Arango 2023). This is also territorialization.

Following the discussion above, we can also begin to understand why many scholars and activists are using the term 'reterritorialization', broadly defined to include cases in which the landless have occupied territories and managed to influenced their state's decision to recognize such a bold and insurgent initiative, but to also include cases in which relatively settled villagers in a territory are suddenly expelled from the land to reallocate resources to new investors. Shattuck and Peluso (2021) also discuss the everyday forms of territorialization that ordinary people undertake outside the official processes of the state. This category can include informal colonization of a portion of national parks and other public spaces, for example.

In the context of Critical Agrarian Studies, the concept of territorialization reminds us that land-as-a-plot-of-farmland, which is often a unit of analysis in the field, is important, but it is not the only relevant way of understanding the politics of land. The concepts of territory and territorialization are important examples in this context, as are our discussions elsewhere in this book on the concepts of 'landscape' and 'land in production/social reproduction'.

Recommended reading

Shattuck, A., & Peluso, N. L. (2021). Territoriality. In H. Akram-Lodhi, K. Dietz, B. Engels, & B. McKay (Eds.), *Handbook of critical agrarian studies* (pp. 197–204). Cheltenham: Edward Elgar Publishing.

Vandergeest, P., & Peluso, N. L. (1995). Territorialization and state power in Thailand. *Theory and Society*, 24(3), 385–426.

W

Wasteland

In the seventeenth century, John Locke pioneered the labour theory of property; that is, he who expends labour on land transforms that land into his private property from which he can exclude others. That idea proved to have an enduring power. It is premised on the assumption that there are two broad types of lands: enclosed and settled lands, and those that are in the commons. The assumption is that lands in the commons are not worked by labour and are considered to be, in his own word, 'wasteland'. The core elements of Locke's idea are summarized below:

> The *Labour* of his Body, and the *Work* of his Hands, we may say, are properly his. Whatsoever then he removes out of the State that Nature hath provided, and left it in, he hath mixed his *Labour* with, and joyned to it something that is his own, and thereby makes it his *Property.* It being by him removed from the common state Nature placed it in, it hath by this *labour* something annexed to it, that excludes the common right of other Men. For this *Labour* being the unquestionable Property of the Labourer, no Man but he can have a right to what that is once joyned to, at least where there is enough, and as good left in common for others.
>
> (Locke 1978: 18, original emphases)

> … So little, that even amongst us, Land that is left wholly to Nature, that hath no improvement of Pasturage, Tillage, or Planting, is called, indeed it is, *wast;* and we shall find the benefit of it amount to little more than nothing.
>
> (Locke 1978: 24, original emphasis)

Locke also talked about the eventual accumulation of vast lands by some landowners with the use of money, as something reasonable and fitting logically with his labour theory.

DOI: 10.4324/9781032658438-16

This idea promulgated by Locke was used extensively in the colonies to declare the commons—including those of Indigenous peoples—as 'wasteland'. It has been a foundational idea that framed the British colonial land and forest policies in India, for example (Whitehead 2012).

In the twenty-first century, Locke's wasteland idea remains a key driving force in recasting the politics of land classification and reallocation. The global land rush that erupted in the 2000s (see 'Land rush', this book) highlighted competing perspectives about land use and reallocation, and Locke's idea remains as influential as it was during colonial times.

In 2012, the Myanmar government enacted a contemporary version of the British colonial era concept of wasteland, with its Vacant, Fallow and Virgin Land Management Law or VFV Law. It is premised on Locke's classic formulation: rural villagers are asked to register their lands, mostly held under customary land tenure, much of it using the swidden agriculture system of production which by western, neoclassical economics standards is considered economically inefficient. Those that registered were given the so-called Form 7; they were then expected, or coerced, by the state to pursue sedentary farming (put differently: they are no longer allowed to pursue mobilize agriculture). All lands not registered were declared to be automatically classified as 'vacant, fallow, and virgin' and assumed to be idle and unused. These lands were seized by the central government and reallocated to big land investors, either military entities, crony capitalists, or foreign investors. Agrarian activists call it the 'land grabbing law' (TNI 2018).

Recommended reading

Locke, J. (1978). Of property. In C. B. Macpherson (Ed.), *Mainstream and critical positions* (pp. 15–28). Toronto: University of Toronto Press.

References

Agrawal, A., Erbaugh, J., & Pradhan, N. (2023). The commons. *Annual Review of Environment and Resources*, 48(1), 531–558.

Agarwal, B. (1994). *A Field of One's Own: Gender and Land Rights in South Asia*. Cambridge: Cambridge University Press.

Aguiar, D., Ahmed, Y., Avcı, D., Bastos, G., Batubara, B., Bejeno, C., Camacho-Benavides, C. I., Chauhan, K., Coronado, S., Das, S., Ejarque, M., Benlisoy, Z. C. E., Güiza-Gómez, D. I., Gyapong, A. Y., Phan, H. P., Hassan, R., Rodríguez, C. H., Ng, H., Hussain, S. B., Kavak, S., Kelegama, T., Kurien, A. J., Leung, D. S., Martínez-Cruz, T., Monjane, B., Mudimu, G. T., Pelek, D., Ralandison, T., Sosa Varrotti, A. P., Torvikey, D., & Valencia-Duarte, D. M. (2023). Transforming critical agrarian studies: Solidarity, scholar-activism and emancipatory agendas in and from the Global South. *The Journal of Peasant Studies*, 50(2), 758–786.

Akram-Lodhi, A. H. (2021). The ties that bind? Agroecology and the agrarian question in the twenty-first century. *The Journal of Peasant Studies*, 48(4), 687–714.

Akram-Lodhi, A. H., Borras Jr, S. M., & Kay, C. (Eds.) (2007). *Land, Poverty and Livelihoods in an Era of Globalization: Perspectives from Developing and Transition Countries*. London: Routledge.

Akram-Lodhi, A. H., Dietz, K., Engels, B., & McKay, B. M. (2021). An introduction to the handbook of critical agrarian studies. In A. Akram Lodhi et al. (Eds.), *Handbook of Critical Agrarian Studies*, pp. 1–7. Cheltenham: Edward Elgar.

Akram-Lodhi, A. H., & Kay, C. (2010a). Surveying the agrarian question (part 1): Unearthing foundations, exploring diversity. *The Journal of Peasant Studies*, 37(1), 177–202.

Akram-Lodhi, A. H., & Kay, C. (2010b). Surveying the agrarian question (part 2): Current debates and beyond. *The Journal of Peasant Studies*, 37(2), 255–284.

Alavi, H., & Shanin, T. (1988). Peasantry and capitalism, Introduction to K. Kautsky (1988). In K. Kautsky (Ed.), *The Agrarian Question*, pp. xi–xxxix. London: Zwan Publications.

Alden Wily, L. (2012). Looking back to see forward: The legal niceties of land theft in land rushes. *The Journal of Peasant Studies*, 39(3–4), 751–775.

Alonso-Fradejas, A. (2012). Land control-grabbing in Guatemala: The political economy of contemporary agrarian change. *Canadian Journal of Development Studies/ Revue Canadienne d'études du Développement*, 33(4), 509–528.

Alonso-Fradejas, A. (2021). 'Leaving no one unscathed' in sustainability transitions: The life purging agro-extractivism of corporate renewables. *Journal of Rural Studies*, 81, 127–138.

Amnesty International (2014). *A Land Title Is Not Enough: Ensuring Sustainable Land Restitution in Colombia*. London: Amnesty International.

Andersson, E., Brogaard, S., & Olsson, L. (2011). The political ecology of land degradation. *Annual Review of Environment and Resources*, 36(1), 295–319.

Andrade, D. (2022). Neoliberal extractivism: Brazil in the twenty-first century. *The Journal of Peasant Studies*, 49(4), 793–816.

Andreas, J., Kale, S. S., Levien, M., & Zhang, Q. F. (2020). Rural land dispossession in China and India. *The Journal of Peasant Studies*, 47(6), 1109–1142.

Anseeuw, W., Lay, J., Messerli, P., Giger, M., & Taylor, M. (2013). Creating a public tool to assess and promote transparency in global land deals: The experience of the Land Matrix. *The Journal of Peasant Studies*, 40(3), 521–530.

Anthias, P. (2021). Rethinking territory and property in indigenous land claims. *Geoforum*, 119, 268–278.

Arango, L. (2023). Indigenous peoples, commons and the challenge of sustaining life amid capitalist land grabs'. *Land Rush Working Paper & Notes* No. 4. https://www.iss.nl/en/research/research-projects/commodity-land-rushes-and-regimes/land-rush-working-paper-notes-series

Arango, L. (2024). Spectacularization and the contemporary land rush in the Colombian Altillanura. *Land Deal Politics Initiative (LDPI) Working Paper Series* No. 2. https://www.peasantjournal.org/news/working-paper-series-international-conference-on-global-land-grabbing-bogota-colombia/. Downloaded 15 September 2024.

Arsel, M., & Büscher, B. (2012). Nature™ Inc.: Changes and continuities in neoliberal conservation and market-based environmental policy. *Development and Change*, 43(1), 53–78.

Ashwood, L., Canfield, J., Fairbairn, M., & De Master, K. (2022). What owns the land: The corporate organization of farmland investment. *The Journal of Peasant Studies*, 49(2), 233–262.

Beban, A., & Gorman, T. (2017). From land grab to agrarian transition? Hybrid trajectories of accumulation and environmental change on the Cambodia–Vietnam border. *The Journal of Peasant Studies*, 44(4), 748–768.

Bellisario, A. (2007). The Chilean agrarian transformation: Agrarian reform and Capitalist 'Partial' counter-agrarian reform, 1964–1980: Part 2: CORA, post-1980 outcomes and the emerging agrarian class structure. *Journal of Agrarian Change*, 7(2), 145–182.

Bernstein, H. (2006). Is there an agrarian question in the 21st century?. *Canadian Journal of Development Studies/Revue Canadienne d'études du Développement*, 27(4), 449–460.

Bernstein, H. (2010). *Class Dynamics of Agrarian Change*. Halifax, NS: Fernwood Publishing.

Binswanger, H. (1996). The political implications of alternative models of land reform and compensation. In J. V. Zyl, J. Kirsten & H. P. Binswanger (Eds.), *Agricultural Land Reform in South Africa: Policies, Markets and Mechanisms*, pp. 139–146. Oxford: Oxford University Press.

Bloch, M. (1992 [orig. 1954]). *The Historian's Craft*. Manchester: Manchester University Press.

Boelens, R., Escobar, A., Bakker, K., Hommes, L., Swyngedouw, E., Hogenboom, B., Huijbens, E. H., Jackson, S., Vos, J., Harris, L. M., Joy, K. J., de Castro, F., Duarte-Abadía, B., Tubino de Souza, D., Lotz-Sisitka, H., Hernández-Mora, N.,

Martínez-Alier, J., Roca-Servat, D., Perreault, T., Sanchis-Ibor, C., Suhardiman, D., Ulloa, A., Wals, A., Hoogesteger, J., Hidalgo-Bastidas, J. P., Roa-Avendaño, T., Veldwisch, G. J., Woodhouse, P., & Wantzen, K. (2023). Riverhood: Political ecologies of socionature commoning and translocal struggles for water justice. *The Journal of Peasant Studies*, 50(3), 1125–1156.

Boone, C. (2013). Land regimes and the structure of politics: Patterns of land-related conflict. *Africa*, 83(1), 188–203.

Borras Jr, S. M. (2003). Questioning market-led agrarian reform: Experiences from Brazil, Colombia and South Africa. *Journal of Agrarian Change*, 3(3), 367–394.

Borras Jr, S. M. (2007). *Pro-poor Land Reform: A Critique*. Ottawa: University of Ottawa Press/Les Presses de l'Université d' Ottawa. Open access ebook: https://library.oapen.org/handle/20.500.12657/33025

Borras Jr, S. M., Arango, L., Belay, M., Franco, J. C., Kham, S. S., Moreda, T., Rojas, I., Wang, C., Ye, J., & Xu, Y. (2024). Death of agrarian societies by a thousand cuts: 'Pin prick' land grabs and the land rush. *LDPI Working Paper Series* No. 43. https://www.peasantjournal.org/. Downloaded 14 August 2024.

Borras Jr, S. M., & Franco, J. C. (2023). *Scholar-activism and Land Struggles*. Rugby: Practical Action Publishing. Open access ebook: https://practicalactionpublishing.com/book/2701/scholar-activism-and-land-struggles

Borras Jr, S. M., & Franco, J. C. (2024a). Land rush. *The Journal of Peasant Studies*, 1–16. https://doi.org/10.1080/03066150.2024.2317961

Borras Jr, S. M., & Franco, J. C. (Eds.) (2024b). *The Oxford Handbook of Land Politics*. Oxford: Oxford University Press.

Borras Jr, S. M., Franco, J. C., Ra, D., Kramer, T., Kamoon, M., Phyu, P., Ju, K. K., Vervest, P., Oo, M., Shell, K. Y., Soe, T. M., Dau, Z., Phyu, M., Poine, M. S., Jumper, M. P., Mon, N. S., Oo, K., Thu, K., Khine, N. K., Naing, T. T., Papa, N., Htwe, L. H., Reang, L. H., Jay, L. P., Jai, N. S., Xu, Y., Wang, C., & Ye, J. (2022). Rurally rooted cross-border migrant workers from Myanmar, Covid-19, and agrarian movements. *Agriculture and Human Values*, 39(1), 315–338.

Borras Jr, S. M., Franco, J. C., Gómez, S., Kay, C., & Spoor, M. (2012). Land grabbing in Latin America and the Caribbean. *The Journal of Peasant Studies*, 39(3–4), 845–872.

Borras Jr, S. M., Franco, J. C., Isakson, S. R., Levidow, L., & Vervest, P. (2016). The rise of flex crops and commodities: Implications for research. *The Journal of Peasant Studies*, 43(1), 93–115.

Borras Jr, S. M., Franco, J. C., & Nam, Z. (2020a). Climate change and land: Insights from Myanmar. *World Development*, 129, 104864.

Borras Jr, S. M., Franco, J. C., & Nam, Z. (2020a). Climate change and land: Insights from Myanmar. *World Development*, 129, 104864.

Borras Jr, S. M., Mills, E. N., Seufert, P., Backes, S., Fyfe, D., Herre, R., & Michéle, L. (2020b). Transnational land investment web: Land grabs, TNCs, and the challenge of global governance. *Globalizations*, 17(4), 608–628.

Borras Jr, S. M. (2023). Politically engaged, pluralist and internationalist: Critical agrarian studies today. *The Journal of Peasant Studies*, 50(2), 449–489.

Brockington, D. (2002). *Fortress Conservation: The Preservation of the Mkomazi Game Reserve, Tanzania*. Bloomington: Indiana University Press.

Brockington, D., & Duffy, R. (Eds.) (2011). *Capitalism and Conservation*. London: John Wiley & Sons.

Bruna, N. (2022). A climate-smart world and the rise of Green Extractivism. *The Journal of Peasant Studies*, 49(4), 839–864.

Burawoy, M. (1976). The functions and reproduction of migrant labor: Comparative material from Southern Africa and the United States. *American Journal of Sociology*, 81(5), 1050–1087.

Burnod, P., Gingembre, M., & Andrianirina Ratsialonana, R. (2013). Competition over authority and access: International land deals in Madagascar. *Development and Change*, 44(2), 357–379.

Büscher, B. (2024). The great conservation tragedy? A critical reflection of (neo) protectionism in relation to the '30×30' global biodiversity framework. *The Journal of Peasant Studies*, 1–22. https://doi.org/10.1080/03066150.2024.2397373

Byres, T. J. (1979). Of neo-populist pipe-dreams: Daedalus in the Third World and the myth of urban bias. *The Journal of Peasant Studies*, 6(2), 210–244.

Byres, T. J. (2004). Neo-classical neo-populism 25 years on: Déjà vu and déjà passé. Towards a critique. *Journal of Agrarian Change*, 4(1–2), 17–44.

Caffentzis, G., & Federici, S. (2014). Commons against and beyond capitalism. *Community Development Journal*, 49(suppl_1), 92–105.

Calmon, D., Jacovetti, C., & Koné, M. (2021). Agrarian climate justice as a progressive alternative to climate security: Mali at the intersection of natural resource conflicts. *Third World Quarterly*, 42(12), 2785–2803.

Castro-Vargas, M. S., & Werner, M. (2023). Regulation by impasse: Pesticide registration, capital and the state in Costa Rica. *Environment and Planning E: Nature and Space*, 6(2), 901–922.

Chagnon, C. W., Durante, F., Gills, B. K., Hagolani-Albov, S. E., Hokkanen, S., Kangasluoma, S. M., Konttinen, H., Kröger, M., LaFleur, W., Ollinaho, O., & Vuola, M. P. (2022). From extractivism to global extractivism: The evolution of an organizing concept. *The Journal of Peasant Studies*, 49(4), 760–792.

Chambati, W. (2024). Land alienation, proletarianization, and changing labor market regimes in Southern Africa. In S. M. Borras & J. C. Franco (Eds.), *The Oxford Handbook of Land Politics*. Oxford University Press. https://doi.org/10.1093/oxfordhb/9780197618646.013.4

Chimhowu, A. (2019). The 'new' African customary land tenure. Characteristic, features and policy implications of a new paradigm. *Land Use Policy*, 81, 897–903.

Chitonge, H. (2022). Resettled but not redressed: Land restitution and post-settlement dynamics in South Africa. *Journal of Agrarian Change*, 22(4), 722–739.

Clapp, J. (2014). Food security and food sovereignty: Getting past the binary. *Dialogues in Human Geography*, 4(2), 206–211.

Clapp, J., & Isakson, S. R. (2018). *Speculative Harvests: Financialization, Food, and Agriculture*. Rugby: Practical Action; Halifax, NS: Fernwood. Open access ebook: https://practicalactionpublishing.com/book/2046/speculative-harvests

Corbera, E. (2012). Problematizing REDD+ as an experiment in payments for ecosystem services. *Current Opinion in Environmental Sustainability*, 4(6), 612–619.

Corbera, E., & Schroeder, H. (2011). Governing and implementing REDD+. *Environmental Science & Policy*, 14(2), 89–99.

Coronado Delgado, S. A. (2022). Peasants, protests and litigation: Struggles over land and institutions in Colombia. The Hague: International Institute of Social Studies, Doctoral dissertation.

Cortesi, L. (2021). An ontology of water and land in North Bihar, India. *Journal of the Royal Anthropological Institute*, 27(4), 870–889.

Cotula, L., Oya, C., Codjoe, E. A., Eid, A., Kakraba-Ampeh, M., Keeley, J., Kidewa, A. L., Makwarimba, M., Seide, W. M., Nasha, W. O., Asare, R. O., & Rizzo, M. (2014).

Testing claims about large land deals in Africa: Findings from a multi-country study. *Journal of Development Studies*, 50(7), 903–925.

Cousins, B. (1997). How do rights become real? Formal and informal institutions in South Africa's land reform. *IDS Bulletin*, 28(4), 59–68.

Cousins, B. (2024). Land, social reproduction, and agrarian change. In S. M. Borras & J. C. Franco (Eds.), *The Oxford Handbook of Land Politics*. Oxford University Press. https://doi.org/10.1093/oxfordhb/9780197618646.013.32

Crippa, M., Solazzo, E., Guizzardi, D., Monforti-Ferrario, F., Tubiello, F. N., & Leip, A. J. N. F. (2021). Food systems are responsible for a third of global anthropogenic GHG emissions. *Nature Food*, 2(3), 198–209.

Cushman, G. T. (2013). *Guano and the Opening of the Pacific World: A Global Ecological History*. Cambridge: Cambridge University Press.

Davis, M. (2006). *Planet of Slums*. New York: Verso.

De Angelis, M. (2013). Does capital need a commons fix?. *Ephemera*, 13(3), 603–615.

De Soto, H. (2000). *The Mystery of Capital: Why Capitalism Triumphs in the West and Fails Everywhere Else*. New York: Basic books.

Deere, C. D. (1985). Rural women and state policy: The Latin American agrarian reform experience. *World Development*, 13(9), 1037–1053.

Deere, C. D. (1995). What difference does gender make? Rethinking peasant studies. *Feminist Economics*, 1(1), 53–72.

Deere, C. D., & De Janvry, A. (1981). Demographic and social differentiation among northern Peruvian peasants. *The Journal of Peasant Studies*, 8(3), 335–366.

Deininger, K. (1999). Making negotiated land reform work: Initial experience from Colombia, Brazil and South Africa. *World Development*, 27(4), 651–672.

Deininger, K. (2011). Challenges posed by the new wave of farmland investment. *The Journal of Peasant Studies*, 38(2), 217–247.

Deininger, K., & Binswanger, H. (1999). The evolution of the World Bank's land policy: Principles, experience, and future challenges. *The World Bank Research Observer*, 14(2), 247–276.

Deininger, K., & Byerlee, D. (2011). *Rising Global Interest in Farmland: Can It Yield Sustainable and Equitable Benefits?*. Washington, DC: World Bank Publications.

Dell'Angelo, J., D'odorico, P., Rulli, M. C., & Marchand, P. (2017). The tragedy of the grabbed commons: Coercion and dispossession in the global land rush. *World Development*, 92, 1–12.

Desmarais, A. A., Qualman, D., Magnan, A., & Wiebe, N. (2015). Land grabbing and land concentration: Mapping changing patterns of farmland ownership in three rural municipalities in Saskatchewan, Canada. *Canadian Food Studies/La Revue Canadienne des études sur l'alimentation*, 2(1), 16–47.

Diskin, M. (1989). El Salvador: Reform prevents change. In W. C. Thiesenhusen (Ed.), *Searching for Agrarian Reform in Latin America*, pp. 429–450. London: Unwin Hyman.

Dressler, W., de Koning, J., Montefrio, M., & Firn, J. (2016). Land sharing not sparing in the "green economy": The role of livelihood bricolage in conservation and development in the Philippines. *Geoforum*, 76, 75–89.

Dunlap, A. (2018). Counterinsurgency for wind energy: The Bíi Hioxo wind park in Juchitán, Mexico. *The Journal of Peasant Studies*, 45(3), 630–652.

Dwyer, M. B. (2015). The formalization fix? Land titling, land concessions and the politics of spatial transparency in Cambodia. *The Journal of Peasant Studies*, 42(5), 903–928.

Dwyer, M. B. (2022). *Upland Geopolitics: Postwar Laos and the Global Land Rush*. Seattle: University of Washington Press.

Edelman, M. (2013). Messy hectares: Questions about the epistemology of land grabbing data. *The Journal of Peasant Studies*, 40(3), 485–501.

Edelman, M. (2021). Hollowed out Heartland, USA: How capital sacrificed communities and paved the way for authoritarian populism. *Journal of Rural Studies*, 82, 505–517.

Edelman, M., & León, A. (2013). Cycles of land grabbing in Central America: An argument for history and a case study in the Bajo Aguán, Honduras. *Third World Quarterly*, 34(9), 1697–1722.

Edelman, M., & Wolford, W. (2017). Introduction: Critical agrarian studies in theory and practice. *Antipode*, 49(4), 959–976.

Fairbairn, M. (2014). 'Like gold with yield': Evolving intersections between farmland and finance. *The Journal of Peasant Studies*, 41(5), 777–795.

Fairbairn, M. (2015). Foreignization, financialization and land grab regulation. *Journal of Agrarian Change*, 15(4), 581–591.

Fairbairn, M. (2020). *Fields of Gold: Financing the Global Land Rush*. Ithaca, NY: Cornell University Press.

Fairhead, J., Leach, M., & Scoones, I. (2012). Green grabbing: A new appropriation of nature?. *The Journal of Peasant Studies*, 39(2), 237–261.

FAO (2013). *Climate-Smart Agriculture Sourcebook*. Rome: FAO.

Fischer, J., Abson, D. J., Butsic, V., Chappell, M. J., Ekroos, J., Hanspach, J., Kuemmerle, T., Smith, H. G., & von Wehrden, H. (2014). Land sparing versus land sharing: Moving forward. *Conservation Letters*, 7(3), 149–157.

Foster, J. B. (1999). Marx's theory of metabolic rift: Classical foundations for environmental sociology. *American Journal of Sociology*, 105(2), 366–405.

Fox, J. (1993). *The Politics of Food in Mexico: State Power and Social Mobilization*. Ithaca, NY: Cornell University Press.

Franco, J. C. (2008). Peripheral justice? Rethinking justice sector reform in the Philippines. *World Development*, 36(10), 1858–1873.

Franco, J. C. (2011). *Bound by Law: Filipino Rural Poor and the Search for Justice in a Plural-legal Landscape*. Quezon City: Ateneo de Manila University Press; Honolulu: University of Hawaii Press.

Franco, J. C., & Borras Jr, S. M. (Eds.) (2005). *On Just Grounds: Struggling for Agrarian Justice and Citizenship Rights in the Rural Philippines*. Quezon City: Institute for Popular Democracy.

Franco, J. C., & Borras Jr, S. M. (2019). Grey areas in green grabbing: Subtle and indirect interconnections between climate change politics and land grabs and their implications for research. *Land Use Policy*, 84, 192–199.

Franco, J. C., & Borras Jr, S. M. (2021). The global climate of land politics. *Globalizations*, 18(7), 1277–1297.

Franco, J. C., & Monsalve, S. (2018). Why wait for the state? Using the CFS Tenure Guidelines to recalibrate political-legal struggles for democratic land control. *Third World Quarterly*, 39(7), 1386–1402.

Franquesa, J. (2018). *Power Struggles: Dignity, Value, and the Renewable Energy Frontier in Spain*. Bloomington: Indiana University Press.

Fraser, A. (2019). Land grab/data grab: Precision agriculture and its new horizons. *The Journal of Peasant Studies*, 46(5), 893–912.

Fraser, N. (2021). Climates of capital. *New Left Review*, (127), 94–127.

Friedmann, H., & McMichael, P. (1989). Agriculture and the state system. *Sociologia Ruralis*, 29(2), 93–117.

Friis, C., & Nielsen, J. Ø. (2016). Small-scale land acquisitions, large-scale implications: Exploring the case of Chinese banana investments in Northern Laos. *Land Use Policy*, 57, 117–129.

Gerber, J. F. (2014). The role of rural indebtedness in the evolution of capitalism. *The Journal of Peasant Studies*, 41(5), 729–747.

Gerber, J. F. (2025). The second contradiction of capitalism. *The Journal of Peasant Studies*. https://doi.org/10.1080/03066150.2025.245128

GRAIN (2008). Seized: The 2008 Land Grab for Food and Financial Security. Barcelona: GRAIN.

Grajales, J. (2021). *Agrarian Capitalism, War and Peace in Colombia: Beyond Dispossession*. London: Routledge.

Green, W. N., & Bylander, M. (2021). The exclusionary power of microfinance: Over-indebtedness and land dispossession in Cambodia. *Sociology of Development*, 7(2), 202–229.

Griffin, K., Khan, A. R., & Ickowitz, A. (2002). Poverty and the distribution of land. *Journal of Agrarian Change*, 2(3), 279–330.

Grindle, M., & Thomas, J. (1989). Policy makers, policy choices, and policy outcomes: The political economy of reform in developing countries. *Policy Sciences*, 22(2), 213–248.

Guthman, J. (2002). Commodified meanings, meaningful commodities: Re–thinking production–consumption links through the organic system of provision. *Sociologia Ruralis*, 42(4), 295–311.

Gyapong, A. Y. (2019). Land deals, wage labour, and everyday politics. *Land*, 8(6), 1–21.

Hall, D. (2011). Land grabs, land control, and Southeast Asian crop booms. *The Journal of Peasant Studies*, 38(4), 837–857.

Hall, D. (2013). Primitive accumulation, accumulation by dispossession and the global land grab. *Third World Quarterly*, 34(9), 1582–1604.

Hall, D. (2023). Russia's invasion of Ukraine and critical agrarian studies. *The Journal of Peasant Studies*, 50(1), 26–46.

Hall, D., Hirsch, P., & Li, T. M. (2011). *Introduction to Powers of Exclusion: Land Dilemmas in Southeast Asia*. Singapore: National University of Singapore Press; Honolulu: University of Hawaii Press.

Hall, R., Edelman, M., Scoones, I., White, B., & Wolford, W. (2015). Resistance, acquiescence or incorporation? *The Journal of Peasant Studies*, 42(3–4), 467–488.

Hall, R., Scoones, I., & Tsikata, D. (2017). Plantations, outgrowers and commercial farming in Africa: Agricultural commercialisation and implications for agrarian change. *The Journal of Peasant Studies*, 44(3), 515–537.

Hardin, G. (1968). The tragedy of the commons. *Science*, 162, 1243–1248.

Hardin, G. (1994). The tragedy of the unmanaged commons. *Trends in Ecology & Evolution*, 9(5), 199.

Harvey, D. (2003). *The New Imperialism*. Oxford: Oxford University Press.

Harvey, D. (2011). The future of the commons. *Radical History Review*, 109, 101–107.

He, J. (2023). Everyday environmental justice in payments for ecosystem services: Insights from two protected areas in China. *Conservation and Society*, 21(3), 153–164.

Heinimann, A., Mertz, O., Frolking, S., Egelund Christensen, A., Hurni, K., Sedano, F., Parsons Chini, L., Sahajpal, R., Hansen, M., & Hurtt, G. (2017). A global view of shifting cultivation: Recent, current, and future extent. *PloS One*, 12(9), e0184479.

Herring, R. (1983). *Land to the Tiller: The Political Economy of Agrarian Reform in South Asia*. New Haven, CT: Yale University Press.

Herring, R. J., & Agarwala, R. (2006). Introduction: Restoring agency to class: Puzzles from the subcontinent. *Critical Asian Studies*, 38(4), 323–356.

Hilhorst, T., Nelen, J., & Traoré, N. (2011). Agrarian change below the radar screen: Rising farmland acquisitions by domestic investors in West Africa. Results from a survey in Benin, Burkina Faso and Niger. Unpublished paper. Amsterdam: Royal Tropical Institute.

Ho, P., & Spoor, M. (2006). Whose land? The political economy of land titling in transitional economies. *Land Use Policy*, 23(4), 580–587.

Hobsbawm, E. J. (1972). The social function of the past: Some questions. *Past & Present*, 55(1), 3–17.

Holleman, H. (2018). *Dust Bowls of Empire*. New Have, CT: Yale University Press.

Holliday, J. S. (1999). *Rush for Riches: Gold Fever and the Making of California*. Berkeley: University of California Press.

Holt Giménez, E., & Shattuck, A. (2011). Food crises, food regimes and food movements: Rumblings of reform or tides of transformation?. *The Journal of Peasant Studies*, 38(1), 109–144.

Hunsberger, C., Corbera, E., Borras Jr, S. M., Franco, J. C., Woods, K., Work, C., de la Rosa, R., Eang, V., Herre, R., Kham, S. S., Park, C., Sokheng, S., Spoor, M., Thein, S., Aung, K. T., Thuon, R., & Vaddhanaphuti, C. (2017). Climate change mitigation, land grabbing and conflict: Towards a landscape-based and collaborative action research agenda. *Canadian Journal of Development Studies/Revue Canadienne d'études du Développement*, 38(3), 305–324.

Hurtado-Hurtado, C., Ortiz-Miranda, D., & Arnalte-Alegre, E. (2023). The legal contention for Baldíos Land in the Colombian Altillanura. *Latin American Politics and Society*, 65(2), 55–82.

IPCC (2019). *Climate Change and Land*. An IPCC special report on climate change, desertification, land degradation, sustainable land management, food security, and greenhouse gas fluxes in terrestrial ecosystems. https://www.ipcc.ch/srccl/. Downloaded 17 May 2020.

Isakson, S. R. (2014). Food and finance: The financial transformation of agro-food supply chains. *The Journal of Peasant Studies*, 41(5), 749–775.

Jacobs, R. (2018). An urban proletariat with peasant characteristics: Land occupations and livestock raising in South Africa. *The Journal of Peasant Studies*, 45(5–6), 884–903.

Jacobs, R. (2024). Land for livelihoods: Urban agriculture and the agrarian question in the 21st century. In S. M. Borras & J. C. Franco (Eds.), *The Oxford Handbook of Land Politics*. Oxford: Oxford University Press.

Jacobs, S. (2009). Gender and land reforms: Comparative perspectives. *Geography Compass*, 3(5), 1675–1687.

Kaag, M. M. A., & Zoomers, A. (2014). *The Global Land Grab: Beyond the Hype*. London: Zed book.

Kandel, M. (2015). Politics from below? Small-, mid-and large-scale land dispossession in Teso, Uganda, and the relevance of scale. *The Journal of Peasant Studies*, 42(3–4), 635–652.

Kautsky, K. (1988). *The Agrarian Question*. London: Zwan Publications. First published in 1899.

Kay, C. (2002). Chile's neoliberal agrarian transformation and the peasantry. *Journal of Agrarian Change*, 2(4), 464–501.

Kay, C. (2009). Development strategies and rural development: Exploring synergies, eradicating poverty. *The Journal of Peasant Studies*, 36(1), 103–137.
Kay, C., & Silva, P. (Eds.) (1992). *Development and Social Change in the Chilean Countryside: From the Pre-land Reform Period to the Democratic Consolidation*. Amsterdam: CEDLA.
Kay, C., & Urioste, M. (2007). Bolivia's unfinished agrarian reform: Rural poverty and development policies. In H. Akram-Lodhi et al. (Eds.), *Land, Poverty and Livelihoods in an Era of Globalization*, pp. 41–79. London: Routledge.
Kay, S. (2014). Reclaiming agricultural investment: Towards public-peasant investment synergies. TNI Agrarian Justice Programme Policy Paper. Amsterdam: Transnational Institute.
Kelly, A. B., & Peluso, N. L. (2015). Frontiers of commodification: State lands and their formalization. *Society & Natural Resources*, 28(5), 473–495.
Kerkvliet, B. (2005). *The Power of Everyday Politics: How Vietnamese Peasants Transformed National Policy*. Ithaca, NY: Cornell University Press.
Kerkvliet, B. (2009). Everyday politics in peasant societies (and ours). *The Journal of Peasant Studies*, 36(1), 227–243.
Kremen, C., Williams, N. M., & Thorp, R. W. (2002). Crop pollination from native bees at risk from agricultural intensification. *Proceedings of the National Academy of Sciences*, 99(26), 16812–16816.
Kröger, M. (2014). The political economy of global tree plantation expansion: A review. *The Journal of Peasant Studies*, 41(2), 235–261.
Kröger, M., & Margutti, A. (2024). Land title frauds as corporate land-grabbing methods: Evidence from Brazil. *Land Deal Politics Initiative (LDPI) Working Paper* No. 4. https://www.peasantjournal.org/news/working-paper-series-international-conference-on-global-land-grabbing-bogota-colombia/. Downloaded 15 September 2024.
Lahiff, E., Borras Jr, S. M., & Kay, C. (2007). Market-led agrarian reform: Policies, performance and prospects. *Third World Quarterly*, 28(8), 1417–1436.
Lamb, V., & Dao, N. (2017). Perceptions and practices of investment: China's hydropower investments in Vietnam and Myanmar. *Canadian Journal of Development Studies/Revue canadienne d'études du développement*, 38(3), 395–413.
Lambin, E. F., Geist, H. J., & Lepers, E. (2003). Dynamics of land-use and land-cover change in tropical regions. *Annual Review of Environment and Resources*, 28(1), 205–241.
Larder, N. (2015). Space for pluralism? Examining the Malibya land grab. *The Journal of Peasant Studies*, 42(3–4), 839–858.
Lavers, T. (2012). 'Land grab' as development strategy? The political economy of agricultural investment in Ethiopia. *The Journal of Peasant Studies*, 39(1), 105–132.
Lay, J., Anseeuw, W., Eckert, S., Flachsbarth, I., Kubitza, C., Nolte, K., & Giger, M. (2021). Taking stock of the global land rush: Few development benefits, many human and environmental risks. Analytical report III. Centre for Development and Environment, University of Bern, and others.
Lenin, V. I. (2004 [orig. 1899]). *The Development of Capitalism in Russia*. Honolulu, HI: University Press of the Pacific.
Lenzerini, F. (2019). Implementation of the UNDRIP around the world: Achievements and future perspectives. The outcome of the work of the ILA Committee on the Implementation of the Rights of Indigenous Peoples. *The International Journal of Human Rights*, 23(1–2), 51–62.
León Araya, A. (2023). Monocrops. *The Journal of Peasant Studies*, 50(3), 797–808.

Levien, M. (2013). Regimes of dispossession: From steel towns to special economic zones. *Development and Change*, 44(2), 381–407.
Levien, M. (2018). *Dispossession without Development: Land Grabs in Neoliberal India*. Oxford: Oxford University Press.
Levien, M. (2021). Coercive rentier networks: "Land Mafias" in neoliberal India. *Sociology of Development*, 7(2), 159–185.
Levien, M., Watts, M., & Yan, H. (2018). Agrarian Marxism. *The Journal of Peasant Studies*, 45(5–6), 853–883.
Li, L., & O'Brien, K. J. (2006). *Rightful Resistance in Rural China*. Cambridge: Cambridge University Press.
Li, T. M. (2010). To make live or let die? Rural dispossession and the protection of surplus populations. *Antipode*, 41, 66–93.
Li, T. M. (2011). Centering labor in the land grab debate. *The Journal of Peasant Studies*, 38(2), 281–298.
Li, T. M. (2014). What is land? Assembling a resource for global investment. *Transactions of the Institute of British Geographers*, 39(4), 589–602.
Li, T. M. (2015). *Land's End: Capitalist Relations on an Indigenous Frontier*. Durham, NC: Duke University Press.
Li, T. M. (2017). Rendering land investible: Five notes on time. *Geoforum*, 82, 272–276.
Locke, J. (1978). Of property. In C. B. Macpherson (Ed.), *Mainstream and Critical Positions*, pp. 15–28. Toronto: University of Toronto Press.
Lund, C. (2006). Twilight institutions: Public authority and local politics in Africa. *Development and Change*, 37(4), 685–705.
Lund, C. (2023). An air of legality–legalization under conditions of rightlessness in Indonesia. *The Journal of Peasant Studies*, 50(4), 1295–1316.
Macpherson, C. B. (Ed.) (1978). *Property, Mainstream and Critical Positions*. Toronto: University of Toronto Press.
Malm, A. (2016). *Fossil Capital: The Rise of Steam Power and the Roots of Global Warming*. New York: Verso books.
Mamdani, M. (2020). *Neither Settler Nor Native: The Making and Unmaking of Permanent Minorities*. Cambridge, MA: The Belknap Press of Harvard University Press.
Martinez-Alier, J. (2021). Mapping ecological distribution conflicts: The EJAtlas. *The Extractive Industries and Society*, 8(4), 100883.
Martinez-Alier, J., Temper, L., Del Bene, D., & Scheidel, A. (2016). Is there a global environmental justice movement?. *The Journal of Peasant Studies*, 43(3), 731–755.
Marx, K. (1976a). *Capital: Volume III*. London: Penguin.
Marx, K. (1976b). *Capital: Volume I*. London: Penguin.
McAfee, K. (2012). The contradictory logic of global ecosystem services markets. *Development and Change*, 43(1), 105–131.
McClintock, N. (2014). Radical, reformist, and garden-variety neoliberal: Coming to terms with urban agriculture's contradictions. *Local Environment*, 19(2), 147–171.
McKay, B. M. (2018). Democratising land control: Towards rights, reform and restitution in post-conflict Colombia. *Canadian Journal of Development Studies/Revue canadienne d'études du développement*, 39(2), 163–181.
McKay, B. M., Fradejas, A. A., & Ezquerro-Cañete, A. (Eds.) (2021). *Agrarian Extractivism in Latin America*. London: Routledge.
McKeon, N. (2017). Are equity and sustainability a likely outcome when foxes and chickens share the same coop? Critiquing the concept of multistakeholder governance of food security. *Globalizations*, 14(3), 379–398.

McMichael, P. (2012). The land grab and corporate food regime restructuring. *The Journal of Peasant Studies*, 39(3–4): 681–701.
McMichael, P. (2013). *Food Regimes and Agrarian Questions*. Halifax: Fernwood Publishing; Rugby: Practical Action. Open access ebook: https://practicalactionpublishing.com/book/770/food-regimes-and-agrarian-questions
Mehta, L., Veldwisch, G. J., & Franco, J. (2012). Introduction to the special issue: Water grabbing? Focus on the (re) appropriation of finite water resources. *Water Alternatives*, 5(2), 193–207.
Mendonça, M. L., & Pitta, F. T. (2022). Land speculation by international financial capital in Brazil. *Latin American Perspectives*, 49(5), 146–160.
Messerli, P., Giger, M., Dwyer, M. B., Breu, T., & Eckert, S. (2014). The geography of large-scale land acquisitions: Analysing socio-ecological patterns of target contexts in the global South. *Applied Geography*, 53, 449–459.
Meyfroidt, P., De Bremond, A., Ryan, C. M., Archer, E., Aspinall, R., Chhabra, A., Camara, G., Corbera, E., DeFries, R., Díaz, S., Dong, J., Ellis, E. C., Erb, K.-H., Fisher, J. A., Garrett, R. D., Golubiewski, N. E., Grau, H. R., Grove, J. M., Haberl, H., Heinimann, A., Hostert, P., Jobbágy, E. G., Kerr, S., Kuemmerle, T., Lambin, E. F., Lavorel, S., Lele, S., Mertz, O., Messerli, P., Metternicht, G., Munroe, D. K., Nagendra, H., Nielsen, J. Ø., Ojima, D. S., Parker, D. C., Pascual, U., Porter, J. R., Ramankutty, N., Reenberg, A., Roy Chowdhury, R., Seto, K. C., Seufert, V., Shibata, H., Thomson, A., Turner II, B. L., Urabe, J., Veldkamp, T., Verburg, P. H., Zeleke, G., & Zu Ermgassen, E. K. (2022). Ten facts about land systems for sustainability. *Proceedings of the National Academy of Sciences*, 119(7), e2109217118.
Mezzadri, A., Stevano, S., Ossome, L., & Bargawi, H. (2024). The social reproduction of agrarian change: Feminist political economy and rural transformations in the global south. An introduction. *Journal of Agrarian Change*, 24(3), e12595. https://doi.org/10.1111/joac.12595
Mintz, S. W. (1986). *Sweetness and Power: The Place of Sugar in Modern History*. London: Penguin.
Mitchell, D. (1996). *The Lie of the Land: Migrant Workers and the California Landscape*. Minneapolis: University of Minnesota Press.
Monjane, B., & Bruna, N. (2020). Confronting agrarian authoritarianism: Dynamics of resistance to PROSAVANA in Mozambique. *The Journal of Peasant Studies*, 47(1), 69–94.
Moore, J. W. (2000). Sugar and the expansion of the early modern world-economy: Commodity frontiers, ecological transformation, and industrialization. *Review (Fernand Braudel Center)*, 23(3), 409–433.
Moore, J. W. (2011). Transcending the metabolic rift: A theory of crises in the capitalist world-ecology. *The Journal of Peasant Studies*, 38(1), 1–46.
Moore, J. W. (2017). The Capitalocene, Part I: On the nature and origins of our ecological crisis. *The Journal of Peasant Studies*, 44(3), 594–630.
Moreda, T. (2015). Listening to their silence? The political reaction of affected communities to large-scale land acquisitions: Insights from Ethiopia. *The Journal of Peasant Studies*, 42(3–4), 517–539.
Moyo, S. (2011). Three decades of agrarian reform in Zimbabwe. *Journal of Peasant Studies*, 38(3), 493–531.
Moyo, S., & Yeros, P. (Eds.) (2005). *Reclaiming the Land: The Resurgence of Rural Movements in Africa, Asia and Latin America*. London: Zed Books.
Mudimu, G. T., Zuo, T., Shah, A. A., Nalwimba, N., & Ado, A. M. (2021). Land leasing in a post-land reform context: Insights from Zimbabwe. *GeoJournal*, 86, 2927–2943.

Neef, A., Ngin, C., Moreda, T., & Mollett, S. (Eds.) (2023). *Routledge Handbook of Global Land and Resource Grabbing*. London: Routledge.
Nolte, K., Chamberlain, W., & Giger, M. (2016). International Land Deals for Agriculture. Fresh insights from the Land Matrix: Analytical Report II (Centre for Development and Environment, University of Bern; Centre de coopération internationale en recherche agronomique pour le développement; German Institute of Global and Area Studies; University of Pretoria; Bern Open Publishing).
Ntsebeza, L., & Hall, R. (Eds.) (2007). *The Land Question in South Africa: The Challenge of Transformation and Redistribution*. Cape Town: HSRC Press.
O'Connor, J. (2017 [orig. 1973]). *The Fiscal Crisis of the State*. London: Routledge.
O'Connor, J. (1998). *Natural Causes: Essays in Ecological Marxism*. New York: Guilford Press.
Ojeda, D. (2012). Green pretexts: Ecotourism, neoliberal conservation and land grabbing in Tayrona National Natural Park, Colombia. *The Journal of Peasant Studies*, 39(2), 357–375.
O'Laughlin, B. (2022). No separate spheres: The contingent reproduction of living labor in Southern Africa. *Review of International Political Economy*, 29(6), 1827–1846.
Ondetti, G. (2016). The social function of property, land rights and social welfare in Brazil. *Land Use Policy*, 50, 29–37.
Ossome, L. (2022). Introduction: The social reproductive question of land contestations in Africa. *African Affairs*, 121(484), e9–e24.
Ostrom, E. (1990). *Governing the Commons*. Cambridge: Cambridge University Press.
Ostrom, E. (2001). The puzzle of counterproductive property rights reforms: A conceptual analysis. In A. de Janvry, G. Gordillo, J. P. Platteau & E. Sadoulet (Eds.), *Access to Land, Rural Poverty, and Public Action*, pp. 129–150. Oxford: Oxford University Press.
Otsuki, K. (2023). Contested values of development: Experiencing commodification of livelihoods through displacement and resettlement in Mozambique. *Environment and Planning A: Economy and Space*. https://doi.org/10.117.0308518X231182431.
Ouma, S. (2016). From financialization to operations of capital: Historicizing and disentangling the finance–farmland-nexus. *Geoforum*, 72, 82–93.
Ouma, S., Johnson, L., & Bigger, P. (2018). Rethinking the financialization of 'nature'. *Environment and Planning A: Economy and Space*, 50(3), 500–511.
Oya, C. (2013). Methodological reflections on 'land grab' databases and the 'land grab' literature 'rush'. *The Journal of Peasant Studies*, 40(3), 503–520.
Paprocki, K. (2019). All that is solid melts into the bay: Anticipatory ruination and climate change adaptation. *Antipode*, 51(1), 295–315.
Paprocki, K. (2022). Anticipatory ruination. *The Journal of Peasant Studies*, 49(7), 1399–1408.
Park, C. M. Y., & White, B. (2017). Gender and generation in Southeast Asian agro-commodity booms. *The Journal of Peasant Studies*, 44(6), 1103–1110.
Patel, R. (2009). Food sovereignty. *The Journal of Peasant Studies*, 36(3), 663–706.
Patel, R., & Moore, J. W. (2017). *A History of the World in Seven Cheap Things: A Guide to Capitalism, Nature, and the Future of the Planet*. Berkeley: University of California Press.
Pattenden, J. (2018). The politics of classes of labour: Fragmentation, reproduction zones and collective action in Karnataka, India. *The Journal of Peasant Studies*, 45(5–6), 1039–1059.
Pattenden, J. (2023). Progressive politics and populism: Classes of labour and rural–urban political sociology – An introduction to the special issue. *Journal of Agrarian Change*, 23(1), 3–21.

Peluso, N. L., & Lund, C. (2011). New frontiers of land control: Introduction. *The Journal of Peasant Studies*, 38(4), 667–681.

Pereira, J. M. M. (2021). The World Bank and market-assisted land reform in Colombia, Brazil, and Guatemala. *Land Use Policy*, 100, 104909.

Perfecto, I., & Vandermeer, J. (2010). The agroecological matrix as alternative to the land-sparing/agriculture intensification model. *Proceedings of the National Academy of Sciences*, 107(13), 5786–5791.

Peters, P. E. (2004). Inequality and social conflict over land in Africa. *Journal of Agrarian Change*, 4(3), 269–314.

Peters, P. E. (2013). Land appropriation, surplus people and a battle over visions of agrarian futures in Africa. *The Journal of Peasant Studies*, 40(3), 537–562.

Phalan, B., Onial, M., Balmford, A., & Green, R. E. (2011). Reconciling food production and biodiversity conservation: Land sharing and land sparing compared. *Science*, 333(6047), 1289–1291.

Polanyi, K. (2001 [orig. 1944]). *The Great Transformation. The Political and Economic Origins of Our Time*. Boston, MA: Beacon Press.

Putzel, J. (1992). *A Captive Land: The Politics of Agrarian Reform in the Philippines*. Quezon City: Ateneo de Manila University Press.

Ra, D., & Ju, K. K. (2021). 'Nothing about us, without us': Reflections on the challenges of building Land in Our Hands, a national land network in Myanmar/Burma. *The Journal of Peasant Studies*, 48(3), 497–516.

Ra, D., Kham, S. S., Barbesgaard, M., Franco, J. C., & Vervest, P. (2021). The politics of Myanmar's agrarian transformation. *The Journal of Peasant Studies*, 48(3), 463–475.

Rasmussen, M. B., & Lund, C. (2018). Reconfiguring frontier spaces: The territorialization of resource control. *World Development*, 101, 388–399.

Razavi, S. (2009). Engendering the political economy of agrarian change. *The Journal of Peasant Studies*, 36(1), 197–226.

Ribot, J. (2014). Cause and response: Vulnerability and climate in the Anthropocene. *The Journal of Peasant Studies*, 41(5), 667–705.

Ribot, J. C., & Peluso, N. L. (2003). A theory of access. *Rural Sociology*, 68(2), 153–181.

Rojas, I. (2022). The land rush in the Colombian Amazon, changing labour dynamics, and the politics of climate change. *Land Rush Working Paper & Notes*. https://www.iss.nl/en/media/2023-10-2023rojaslimits-and-possibilities-rrushes-5-working-paper-final. Downloaded 15 September 2024.

Rojas, I. (2023). Land is life, land is hope, land is peace: The history of the resettlement of peace signatories in Caquetá, Colombia. *Land & Life Stories Series # 2*. The Hague: International Institute of Social Studies (ISS) or Erasmus University Rotterdam, RRUSHES-5 Program. https://www.iss.nl/en/research/research-projects/commodity-land-rushes-and-regimes/land-life-story-series. Downloaded 3 November 2024.

Rojas, I. (2025). Limits and possibilities of contemporary land struggles by Indigenous Peoples, Black Communities and Campesinxs in the Colombian Amazon. *The Journal of Peasant Studies*, 1–30. https://doi.org/10.1080/03066150.2024.2435557

Roman-Alcalá, A. (2015). Broadening the land question in food sovereignty to northern settings: A case study of occupy the farm. *Globalizations*, 12(4), 545–558.

Rosset, P. M., & Altieri, M. A. (2017). *Agroecology: Science and Politics*. Rugby: Practical Action.

Sack, R. D. (1986). *Human Territoriality: Its Theory and History*. Cambridge: Cambridge University Press.

Saito, K. (2023). *Marx in the Anthropocene: Towards the Idea of Degrowth Communism*. Cambridge: Cambridge University Press.

Salerno, T. (2014). Capitalising on the financialisation of agriculture: Cargill's land investment techniques in the Philippines. *Third World Quarterly*, 35(9), 1709–1727.
Salerno, T. (2017). Cargill's corporate growth in times of crises: How agro-commodity traders are increasing profits in the midst of volatility. *Agriculture and Human Values*, 34, 211–222.
Sanderson, S. W. (1984). *Land Reform in Mexico: 1920–1980*. New York: Academic Press.
Sauer, S., & Pereira Leite, S. (2012). Agrarian structure, foreign investment in land, and land prices in Brazil. *The Journal of Peasant Studies*, 39(3–4), 873–898.
Scheidel, A., & Work, C. (2018). Forest plantations and climate change discourses: New powers of 'green' grabbing in Cambodia. *Land Use Policy*, 77, 9–18.
Scoones, I. (2020). Pastoralists and peasants: Perspectives on agrarian change. *The Journal of Peasant Studies*, 48(1), 1–47.
Scoones, I., Edelman, M., Borras Jr, S. M., Hall, R., Wolford, W., & White, B. (2018). Emancipatory rural politics: Confronting authoritarian populism. *The Journal of Peasant Studies*, 45(1), 1–20.
Scoones, I., Hall, R., Borras Jr, S. M., White, B., & Wolford, W. (2013). The politics of evidence: Methodologies for understanding the global land rush. *The Journal of Peasant Studies*, 40(3), 469–483.
Scoones, I., Marongwe, N., Mavedzenge, B., Murimbarimba, F., Mahenehene, J., & Sukume, C. (2011). Zimbabwe's land reform: Challenging the myths. *The Journal of Peasant Studies*, 38(5), 967–993.
Scoones, I., Smalley, R., Hall, R., & Tsikata, D. (2019). Narratives of scarcity: Framing the global land rush. *Geoforum*, 101, 231–241.
Scott, J. C. (1972). The erosion of patron-client bonds and social change in rural Southeast Asia. *The Journal of Asian Studies*, 32(1), 5–37.
Scott, J. C. (1976). *The Moral Economy of the Peasant: Rebellion and Subsistence in Southeast Asia*. New Haven, CT: Yale University Press.
Scott, J. C. (1998). *Seeing Like a State: How Certain Schemes to Improve the Human Condition Have Failed*. New Haven, CT: Yale University Press.
Sekine, Y. (2021). Emerging 'agrarian climate justice' struggles in Myanmar. *The Journal of Peasant Studies*, 48(3), 517–540.
Serrano, A. (2023). Claiming the right to the city beyond the city: The role of agrarian social movements. *Space and Polity*, 27(2), 234–251.
Seto, K. C., Reenberg, A., Boone, C. G., Fragkias, M., Haase, D., Langanke, T., Marcotullio, P., Munroe, D. K., Olah, B., & Simon, D. (2012). Urban land teleconnections and sustainability. *Proceedings of the National Academy of Sciences*, 109(20), 7687–7692.
Shah, A., & Lerche, J. (2020). Migration and the invisible economies of care: Production, social reproduction and seasonal migrant labour in India. *Transactions of the Institute of British Geographers*, 45(4), 719–734.
Shankland, A., & Gonçalves, E. (2016). Imagining agricultural development in south–south cooperation: The contestation and transformation of ProSAVANA. *World Development*, 81, 35–46.
Shattuck, A., Grajales, J., Jacobs, R., Sauer, S., Galvin, S. S., & Hall, R. (2023). Life on the land: New lives for agrarian questions. *The Journal of Peasant Studies*, 50(2), 490–518.
Shattuck, A., & Peluso, N. L. (2021). Territoriality. In H. Akram-Lodho et al. (Eds.), *Handbook of Critical Agrarian Studies*, pp. 197–204. Cheltenham: Edward Elgar Publishing.

Shivji, I. G. (2017). The concept of 'working people'. *Agrarian South: Journal of Political Economy*, 6(1), 1–13.
Siebert, A. (2020). Transforming urban food systems in South Africa: Unfolding food sovereignty in the city. *The Journal of Peasant Studies*, 47(2), 401–419.
Sikor, T. (2006). Politics of rural land registration in post-socialist societies: Contested titling in villages of northwest Vietnam. *Land Use Policy*, 23, 617–628.
Singh, N. (2009). Fighting rural poverty, inequality and low productivity through legal empowerment of the poor. *The Journal of Peasant Studies*, 36(4), 871–892.
Sjaastad, E., & Cousins, B. (2009). Formalisation of land rights in the South: An overview. *Land Use Policy*, 26(1), 1–9.
Sosa Varrotti, A. P., & Gras, C. (2021). Network companies, land grabbing, and financialization in South America. *Globalizations*, 18(3), 482–497.
Soto Hernandez, D., & Newell, P. (2022). Oro blanco: Assembling extractivism in the lithium triangle. *The Journal of Peasant Studies*, 49(5), 945–968.
Spoor, M. (Ed.) (2008). *The Political Economy of Rural Livelihoods in Transition Economies*. London: Routledge.
Springate-Baginski, O., & Kamoon, M. (2021). Defending Shan State's customary tenure systems from below through collective action research. *The Journal of Peasant Studies*, 48(3), 541–559.
Steinmo, S., Thelen, K., & Longstreth, F. (Eds.) (1992). *Structuring Politics: Historical Institutionalism in Comparative Analysis*. Cambridge: Cambridge University Press.
Stock, R., & Birkenholtz, T. (2021). The sun and the scythe: Energy dispossessions and the agrarian question of labor in solar parks. *The Journal of Peasant Studies*, 48(5), 984–1007.
Sud, N. (2014). The men in the middle: A missing dimension in global land deals. *The Journal of Peasant Studies*, 41(4), 593–612.
Suhardiman, D., Giordano, M., & Molle, F. (2012). Scalar disconnect: The logic of transboundary water governance in the Mekong. *Society & Natural Resources*, 25(6), 572–586.
Suhardiman, D., Keovilignavong, O., & Kenney-Lazar, M. (2019). The territorial politics of land use planning in Laos. *Land Use Policy*, 83, 346–356.
Swinburn, B. A., Kraak, V. I., Allender, S., Atkins, V. J., Baker, P. I., Bogard, J. R., Brinsden, H., Calvillo, A., De Schutter, O., Devarajan, R., Ezzati, M., Friel, S., Goenka, S., Hammond, R. A., Hastings, G., Hawkes, C., Herrero, M., Hovmand, P. S., Howden, M., Jaacks, L. M., Kapetanaki, A. B., Kasman, M., Kuhnlein, H. V., Kumanyika, S. K., Larijani, B., Lobstein, T., Long, M. W., Matsudo, V. K. R., Mills, S. D. H., Morgan, G., Morshed, A., Nece, P. M., Pan, A., Patterson, D. W., Sacks, G., Shekar, M., Simmons, G. L., Smit, W., Tootee, A., Vandevijvere, S., Waterlander, W. E., Wolfenden, L., & Dietz, W. H. (2019). The global syndemic of obesity, undernutrition, and climate change: The Lancet Commission report. *The Lancet*, 393(10173), 791–846.
Tarrow, S. (1994). *Power in Movement: Social Movements, Collective Action and Politics*. Cambridge: Cambridge University Press.
Thompson, E. P. (1991 [orig. 1963]). *The Making of the English Working Class*. London: Penguin.
TNI (2018). *A Declaration of War on Us": The 2018 VFV Law Amendment and Its Impact on Ethnic Nationalities*. Amsterdam: Transnational Institute. https://www.tni.org/en/article/a-declaration-of-war-on-us-the-2018-vfv-law-amendment-and-its-impact-on-ethnic. Downloaded 26 August 2024.

Torres Contreras, G. A. (2022). Twenty-five years under the wind turbines in La Venta, Mexico: Social difference, land control and agrarian change. *The Journal of Peasant Studies*, 49(4), 865–883.

Tramel, S. (2024). Territory grabbing and the undermining of Palestinian food sovereignty. *Land Deal Politics Initiative (LDPI) Working Paper Series* No. 12. https://www.peasantjournal.org/news/working-paper-series-international-conference-on-global-land-grabbing-bogota-colombia/. Downloaded 14 September 2024.

Tsikata, D. (2003). Securing women's interests within land tenure reforms: Recent debates in Tanzania. *Journal of Agrarian Change*, 3(1–2), 149–183.

Tsing, A. L. (2000). Inside the economy of appearances. *Public Culture*, 12(1), 115–144.

Tsing, A. L. (2002). Land as law: Negotiating the meaning of property in Indonesia. In F. Richards (Ed.), *Land, Property, and the Environment*, pp. 94–137. Oakland, CA: Institute for Contemporary Studies.

Tsing, A. L., Mathews, A. S., & Bubandt, N. (2019). Patchy anthropocene: Landscape structure, multispecies history, and the retooling of anthropology: An introduction to supplement 20. *Current Anthropology*, 60(S20), S186–S197.

Tuma, E. (1965). *Twenty-Six Centuries of Agrarian Reform: A Comparative Analysis*. Berkeley: University of California Press.

Vaccaro, I., & Beltran, O. (2019). What do we mean by "the commons?" An examination of conceptual blurring over time. *Human Ecology*, 47(3), 331–340.

van der Ploeg, J. D. (2021). The political economy of agroecology. *The Journal of Peasant Studies*, 48(2), 274–297.

van der Ploeg, J. D., Franco, J. C., & Borras Jr, S. M. (2015). Land concentration and land grabbing in Europe: A preliminary analysis. *Canadian Journal of Development Studies/Revue canadienne d'études du développement*, 36(2), 147–162.

van Schalkwyk, H. D., & van Zyl, J. (1996). The Land Market. In J. v. Zyl, J. Kirsten & H. P. Binswanger (Eds.), *Agricultural Land Reform in South Africa: Policies, Markets and Mechanisms*, pp. 310–335. Oxford: Oxford University Press.

van Zyl, J., & Binswanger, H. (1996). Market-assisted land reform: How will it work?. In J. v. Zyl, J. Kirsten & H. P. Binswanger (Eds.), *Agricultural Land Reform in South Africa: Policies, Markets and Mechanisms*, pp. 413–422. Oxford: Oxford University Press.

Vandergeest, P., & Peluso, N. L. (1995). Territorialization and State Power in Thailand. *Theory and Society*, 24(3), 385–426.

Vandergeest, P., & Peluso, N. L. (2015). Political forests. In R. Bryant (Ed.), *The International Handbook of Political Ecology*, pp. 162–175. Northampton, MA: Edward Elgar.

Vandermeer, J., & Perfecto, I. (2024). Land and agroecology: Interpenetrating theses. In S. M. Borras Jr. & J. C. Franco (Eds.), *The Oxford Handbook of Land Politics*. Oxford: Oxford University Press. https://doi.org/10.1093/oxfordhb/9780197618646.013.19

Vermeulen, S., & Cotula, L. (2010). Over the heads of local people: Consultation, consent, and recompense in large-scale land deals for biofuels projects in Africa. *The Journal of Peasant Studies*, 37(4), 899–916.

Vidal, A. (2004). *Conflicting Laws, Overlapping Claims: The Politics of Indigenous Peoples' Land Rights in Mindanao*. Davao: AFRIM.

Vidal, J. (2008). The great green land grab. *The Guardian*, 13 February. https://www.theguardian.com/environment/2008/feb/13/conservation. Downloaded 15 September 2024.

Visser, O. (2017). Running out of farmland? Investment discourses, unstable land values and the sluggishness of asset making. *Agriculture and Human Values*, 34, 185–198.

Visser, O., Clapp, J., & Isakson, S. R. (2015). Introduction to a symposium on global finance and the agri-food sector: Risk and regulation. *Journal of Agrarian Change*, 15(4), 541–548.

Visser, O., Sippel, S. R., & Thiemann, L. (2021). Imprecision farming? Examining the (in) accuracy and risks of digital agriculture. *Journal of Rural Studies*, 86, 623–632.

Walker, C. (2006). Delivery and disarray: The multiple meanings of land restitution. In S. Buhlungu et al. (Eds.), *State of the Nation – South Africa 2006–2006*, pp. 67–92. Cape Town: HSRC Press.

Weis, T. (2010). The accelerating biophysical contradictions of industrial capitalist agriculture. *Journal of Agrarian Change*, 10(3), 315–341.

Weis, T. (2024). Land, industrial livestock, and interspecies relations: The pursuit of scale and the deceits of productivity. In S. M. Borras Jr. & J. C. Franco (Eds.), *The Oxford Handbook of Land Politics*. Oxford: Oxford University Press. https://doi.org/10.1093/oxfordhb/9780197618646.013.31

White, B. (1989). Problems in the empirical analysis of agrarian differentiation. In G. Hart, A. Turton, & B. White (Eds.), *Agrarian Transformations: Local Processes and the State in Southeast Asia*, pp. 15–30. Berkeley: University of California Press.

White, B. (2018). Marx and Chayanov at the margins: Understanding agrarian change in Java. *The Journal of Peasant Studies*, 45(5–6), 1108–1126.

White, B. (2020). *Agriculture and the Generation Problem*. Rugby: Practical Action Publishing. Open access ebook: https://practicalactionpublishing.com/book/45/agriculture-and-the-generation-problem

White, B., Borras Jr, S. M., Hall, R., Scoones, I., & Wolford, W. (2012). The new enclosures: Critical perspectives on corporate land deals. *The Journal of Peasant Studies*, 39(3–4), 619–647.

Whitehead, J. (2012). John Locke, accumulation by dispossession and the governance of colonial India. *Journal of Contemporary Asia*, 42(1), 1–21.

Wolf, E. R. (1969). *Peasant Wars of the Twentieth Century*. Norman: University of Oklahoma Press.

Wolford, W. (2010). *This Land Is Ours Now: Social Mobilization and the Meanings of Land in Brazil*. Durham, NC: Duke University Press.

Wolford, W. (2021). The Plantationocene: A lusotropical contribution to the theory. *Annals of the American Association of Geographers*, 111(6), 1622–1639.

Wolford, W., Borras Jr, S. M., Hall, R., Scoones, I., & White, B. (2013). Governing global land deals: The role of the state in the rush for land. *Development and Change*, 44(2), 189–210.

Wolford, W. W., White, B., Scoones, I., Hall, R., Edelman, M., & Borras, S. M. (2024). Global land deals: What has been done, what has changed, and what's next?. *The Journal of Peasant Studies*, 1–38. https://doi.org/10.1080/03066150.2024.2325685

Woods, K. (2020). Smaller-scale land grabs and accumulation from below: Violence, coercion and consent in spatially uneven agrarian change in Shan State, Myanmar. *World Development*, 127, 104780.

World Bank (2003). *Land Policies for Growth and Poverty Reduction*. Washington, DC: World Bank.

World Bank (2016). *World Bank Group Climate Change Action Plan*. Washington, DC: World Bank.

Wright, E. O. (2016). Socialism and real utopias. In R. Hahnel & E. O. Wright (Eds.), *Alternatives to Capitalism: Proposals for a Democratic Economy*, pp. 25–40. New York: Verso.

Xu, Y. (2018). Land grabbing by villagers? Insights from intimate land grabbing in the rise of industrial tree plantation sector in Guangxi, China. *Geoforum*, 96, 141–149.

Xu, Y., & Borras Jr, S. M. (2024). China and global land use change. In S. M. Borras & J. C. Franco (Eds.), *The Oxford Handbook of Land Politics*. Oxford: Oxford University Press. https://doi.org/10.1093/oxfordhb/9780197618646.013.49

Yaşın, Z. T. (2022). The environmentalization of the agrarian question and the agrarianization of the climate justice movement. *The Journal of Peasant Studies*, 49(7), 1355–1386.

Ye, J. (2015). Land transfer and the pursuit of agricultural modernization in China. *Journal of Agrarian Change*, 15(3), 314–337.

Ye, J., van Der Ploeg, J. D., Schneider, S., & Shanin, T. (2020). The incursions of extractivism: Moving from dispersed places to global capitalism. *The Journal of Peasant Studies*, 47(1), 155–183.

Yeros, P. (2023). Generalized semiproletarianization in Africa. *The Indian Economic Journal*, 71(1), 162–186.

Zhan, S. (2019). Accumulation by and without dispossession: Rural land use, land expropriation, and livelihood implications in China. *Journal of Agrarian Change*, 19(3), 447–464.

Zoomers, A. (2010). Globalisation and the foreignisation of space: Seven processes driving the current global land grab. *The Journal of Peasant Studies*, 37(2), 429–447.

Zoomers, A., & van der Haar, G. (2000). *Current Land Policy in Latin America: Regulating Land Tenure under Neo-liberalism*. Amsterdam: KIT.

Index

MIX

Papier | Fördert gute Waldnutzung

FSC® C083411

Kolibri 360 GmbH
Ferdinand-Jühlke-Straße 7
99095 Erfurt, Deutschland
produktsicherheit@kolibri360.de